Design Technology in Metal and Plastics

By the same author
Metalwork Technology

D
in

G.

Examir
for the

Examir
Southe

Membe

John Murray Albemarle Street London

Acknowledgements

Thanks are due to the following for permission to use the material indicated:

Edward de Bono: extract from *Lateral Thinking: A Text Book of Creativity* (Jonathan Cape) (p. vii).

A. P. P. Durant: examples of Technical Graphics from Hillsea Comprehensive School, Portsmouth (pp. 112–13).

The Plastics and Rubber Institute: main text of booklet *Plastics in Schools – Safety and Hazards* (pp. 134–9).

Examining Boards as below: questions (pp. 126–33).
  Associated Examining Board (*AEB*)
  East Anglian Examinations Board (*EAEB*)
  South East Regional Examinations Board (*SEREB*)
  Southern Regional Examinations Board (*SREB*)

Printed and bound in Hong Kong by Wing King Tong Co Ltd

ISBN 0 7195 3637 5

Contents

Foreword

Extract from *Lateral Thinking: A Text Book of Creativity* by Edward de Bono (Jonathan Cape).

'There is about creativity a mystique of talent and intangibles. This may be justified in the art world where creativity involves aesthetic sensibility, emotional resonance and a gift for expression. But it is not justified outside that world. More and more creativity is coming to be valued as the essential ingredient in change and in progress. It is coming to be valued above knowledge and above technique since both these are becoming so accessible. In order to be able to use creativity one must rid it of this aura of mystique and regard it as a way of using the mind – a way of handling information. This is what lateral thinking is about.

'Lateral thinking is quite distinct from vertical thinking which is the traditional type of thinking. In vertical thinking one moves forward by sequential steps each of which must be justified. The distinction between the two sorts of thinking is sharp. For instance, in lateral thinking one uses information not for its own sake but for its effect. In lateral thinking one may have to be wrong at some stage in order to achieve a correct solution; in vertical thinking (logic or mathematics) this would be impossible. In lateral thinking one may deliberately seek out irrelevant information; in vertical thinking one selects out only what is relevant.

'Lateral thinking is not a substitute for vertical thinking. Both are required. They are complementary. Lateral thinking is generative. Vertical thinking is selective.

'The purpose of thinking is to collect information and to make the best possible use of it. Because of the way the mind works to create fixed concept patterns we cannot make the best use of new information unless we have some means for restructuring the old patterns and bringing them up to date. Our traditional methods of thinking teach us how to refine such patterns and establish their validity. But we shall always make less than the best use of available information unless we know how to create new patterns and escape from the dominance of the old ones. Vertical thinking is concerned with proving or developing concept patterns. Lateral thinking is concerned with restructuring such patterns (insight) and provoking new ones (creativity). Lateral and vertical thinking are complementary. Skill in both is necessary. Yet the emphasis in education has always been exclusively on vertical thinking.'

Introduction

There are many books that describe in detail how teacher and student can produce functional items using a wide variety of materials and techniques. Some of these are excellent for their purpose. Given a set task, a specified material, a preconceived design and a detailed construction guide, a student can produce work of a technically high standard in which he may take a great deal of pride. He is, however, often unaware that the skills he has developed – important as they are – are purely practical ones, that the designs are stereotyped and the end-products themselves often of dubious value. In the future, faced with similar design problems, he is likely to fall back on what he knows already. He may perhaps add an individual flourish here and there but will produce nonetheless an essentially unoriginal piece of work. The reason for this is clear: he has not been taught to *think* for himself.

It is on this point that modern craft teaching parts company – in a very fundamental way – with that hallowed by tradition, and it is on the principle that a design-based course, firmly geared to problem-solving and creative thinking, will clear the path towards a richer, more rewarding and ultimately more useful activity for the student that this book has been written.

What then is creative design? In talking about 'creativity' – particularly in the pure arts – it is easy to lose oneself in the mists of philosophy. In craftwork, however, the issue is clear: to design creatively means to go back to square one and think through the entire project for oneself.

A design-based course seeks to provide the training and conditions that will enable the student to do just this. He is faced with a problem which he has to solve in its entirety. Certain limitations are prescribed – in schools, particularly, materials, tools and time are restricted. The student will also introduce his own inherent limitations: lack of experience or of imagination. But such limitations are if anything a challenge, a spur to creative thinking. They simplify the problem and define its boundaries, so that the student can get his 'intellectual teeth' into what remains.

The basic problem can be stated simply: how to produce an article that fulfils the desired function, is aesthetically pleasing, and makes use of the materials and tools available. Having been given this task, the student is on his own. He must think, solve and create. He must analyse the problem himself, visualize the end-product, consider various ideas, reject some, adopt others and synthesize them into a feasible design. He plans the details of construction and works through to the final stage.

Then, having finished the job, he evaluates it – a vital part of the process. If it does not satisfy him he back-tracks on his thinking – or on his craft skills – to find where he went wrong, and this evaluation gives meaning to everything he has done. He will understand what the design approach is all about. Next time he will be more careful, will practise his technical skills and will produce a better piece of work. What is more, he will have had a rewarding experience. He will not have blindly followed step-by-step instructions and made the same article as everyone else; he will have done it all himself.

A design-based approach to craftwork, besides being a stimulus to the student at the time, has an educational value that goes beyond the craft itself. Designing is a satisfying and worthwhile intellectual activity in its own right. It covers everything from the original conception to the finished article: the function and form, the texture and pattern, the shape, the aesthetic appeal, the colour, the 'feel', the construction, the use of tools and materials – right through to the final appraisal. The student learns that appearance, though important, counts for little if the function is not fulfilled; that these factors too should not be swamped by technical brilliance; that there is a way out of the strictures imposed by a conventional run-of-the-mill approach; that difficult though it may seem at first – and it is difficult, even for experienced designers – it is possible to be original and technically sound at the same time.

He will learn to view the making of the article from a new angle – even commonplace articles such as a table lamp. The standard design of the table lamp is derived from that of the paraffin lamp – a more-

or-less bulbous base to hold the paraffin, a stem for the feed and a globe to guard the flame. Once he considers the problem from basic principles the student will learn that such a tradition is not necessary: a table lamp design can be considered afresh in its entirety, provided that it fulfils its function of supporting a light source at a given height above a plane surface. (This particular example is considered in more detail in Chapter 5.)

In modern design, new materials have a vital part to play. They demand and benefit from new techniques, and knowledge of how these techniques can be used most effectively is necessary to original creative thinking. For example, when acetate sheeting was first introduced, metalwork techniques were used which were not suited to the material. Plastics became synonymous with shoddy workmanship until new techniques were applied and the full potential of the material was realized. The student should learn as much as he can about the different materials available to him.

Similarly, a thorough grasp of technical skills remains as important as ever it was. Standards in this area should not be allowed to deteriorate. It is no use the student having a fine idea if he is unable to carry it through. The fact that it is his design that he is working on can indeed be a spur to mastering the basic skills: he has a proprietory interest in the finished product.

Articles that have a purely decorative function require a rather different approach from those whose function is more practical. There is scarcely any limit to design ideas for a brooch, for example. Here creativity can be given free rein, though frequently finer techniques are necessary to produce a perfect article.

In all fields of craft design there is challenge. This gives a sense of excitement at the start of a prob-lem, lacking in the traditional approach, and a feeling of deep satisfaction when the project turns out well. These are very desirable qualities in any teaching programme.

This book is divided into two main parts. Chapters 1 to 4 provide the student with design problems related to puzzles or other simple designs in which the practical requirements are easy to grasp and the techniques are not too demanding. These problems increase both in complexity and range of techniques. At first some solutions are given to help lead the pupil into the right way of thinking and also to provide guidelines for inexperienced teachers, but later these are dropped except for an occasional hint. If the teacher gives solutions himself, much of the educational value of the approach is lost.

Chapters 5 to 9 present a variety of ideas for fitments and pieces of furniture. Outlines only are given; there are no working drawings. Those students who show a marked ability to conceive fresh ideas should not necessarily follow even these outlines. They can forge ahead with their own ideas, provided that the teacher is satisfied they are worthwhile and practical. Most students, however, will benefit from such starting ideas as are given and will gradually develop a freer, more individual approach as the course progresses. A common fault at this stage is that ideas outrun technique and some students find it extremely difficult to solve the constructional problems inherent in a design. To help overcome this, some solutions to practical problems are provided.

It is hoped therefore that this book will serve two purposes: to provide a series of fresh ideas for a progressive design course for both experienced and inexperienced teachers and to serve as a reference book of design ideas and constructional solutions for students.

1 Simple practical design and problem solving

INTRODUCTORY WORK

Much of the introductory work, such as puzzles, is best solved by drawing the shapes on paper first, cutting them out accurately and using them as patterns. To fix the paper to Perspex or metal use cow gum or a P V A adhesive thinned down with a little water. To finish the edges of Perspex, rub down with emery cloth or wet and dry paper and polish with Perspex polish or with a car cellulose cutting compound and metal polish.

The introductory problems are laid out in the following way:

1 Statement of the problem.
2 Numbered step-by-step solutions to the problem.
3 Accompanying line diagrams.

This enables the work to be attempted in one of two ways. The problem may be handed to the pupil, the solution being retained as a help to the teacher only; or the pupil can be given the step-by-step solutions. Educationally the first method is desirable.

PUZZLES

Squares and triangles

Design a puzzle to be made from thin plastic sheet, e.g. Perspex. The puzzle must be based on the fitting together of parts to form either a square or an equilateral triangle.

1 On paper draw either a square or an equilateral triangle with sides 60 mm long.

2 By drawing two straight lines, divide the area into three shapes that form an interesting pattern. Avoid acute angles that would make the pieces too weak.

3 Cut out the paper shapes and stick them to pieces of plastic sheet. Use different colours.

4 File the pieces of plastic to the size and shape of the paper patterns.

5 Assemble the pieces together and stick them to a thin card backing, trimming it to shape. This will hold the pieces together so that the edges can be trued up.

6 File a piece of plastic sheet to a size of 110 mm square. Centre the assembled pieces on the sheet and mark round with a fine pencil or a scriber. Remove the central area with drills and tension file or junior hacksaw. File to a fit.

7 A backing piece of thin plastic sheet glued in position will improve the puzzle.

8 Separate the three coloured pieces and remove the paper patterns and card backings. The puzzle is to fit the pieces into the cut-out shape.

Some examples of triangular shapes using two cuts and three cuts are shown opposite.

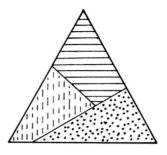

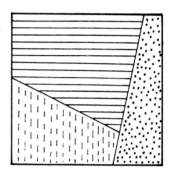

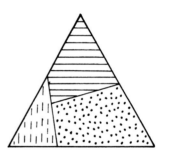

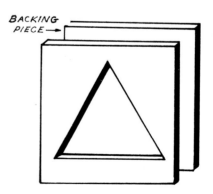

BACKING PIECE →

SOME TRIANGLE PUZZLES

Two cuts

Three cuts

3

Letter shapes

A puzzle can look simple and yet not be simple to solve. Working on this principle, make an assembly puzzle out of thin plastic sheet, the puzzle, when assembled, to form a letter of the alphabet.

1 Draw a block letter to the sizes shown. It is easier to keep to letters that are formed by straight lines.

2 Divide the letter into a number of parts using straight lines. Try to form a key shape as shown in the drawings. Avoid having too many parts and avoid having pieces with acute angles that would lack strength.

3 Carefully cut out the paper patterns and stick them individually to thin plastic sheeting, e.g. Perspex, with a suitable adhesive.

4 File the pieces to size and shape and remove the paper patterns.

5 Clean the edges with wet and dry and polish with Perspex polish or Brasso.

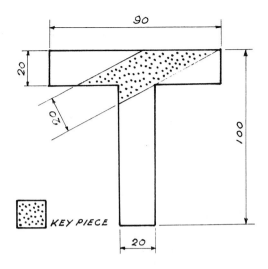

KEY PIECE

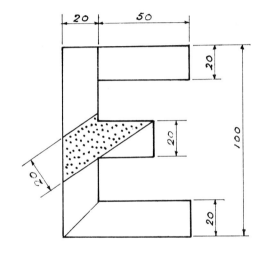

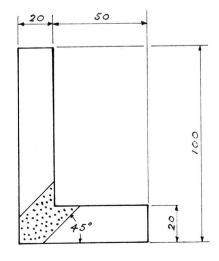

4

Xmas tree

Make a drawing of a Xmas tree decorated with presents. Simplify the drawing and use this to make an assembly puzzle in thin plastic sheet.

1 Make a drawing of a Xmas tree decorated with presents.
2 On a piece of paper 100 mm×80 mm, simplify this drawing allowing a suitable border around it. The drawing must be as simple as possible provided that it symbolizes a Xmas tree with presents.
3 Cut out all the pieces and stick them to thin plastic sheeting, e.g. Perspex, using a suitable adhesive.
4 File the pieces to shape and polish the edges. Remove the paper patterns.
5 Cut and file square a thin piece of plastic to a size of 100 mm×80 mm.
6 Place the pieces of the tree in position on this sheet and mark round with a sharp pencil.
7 Drill and file the internal shapes so that the pieces fit.
8 A piece of thin plastic sheet applied to the back will improve the puzzle.

PICTORIAL

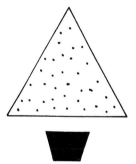

SCHEMATIC

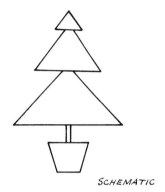

GRAPHIC

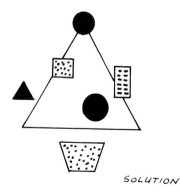

SOLUTION

5

VISUAL SYMBOLS

Visual symbols are now widely used to communicate information in a form that has no language barriers. Some examples are shown opposite.

Make a collection of well-known visual symbols and then design a set of visual symbols that might be used throughout the school to indicate to visitors the various teaching areas.

1 Make a symbol to indicate a school teaching area, e.g. geography, mathematics.

2 Simplify the symbol and keep it bold. The symbol can be formed by fitting parts together or superimposing parts.

3 Draw the symbol within a circle of 70 mm diameter.

4 Cut out the paper parts and stick them to Perspex sheet with a suitable adhesive.

5 File the parts to shape and polish them with Perspex polish. Remove the paper patterns.

6 Cut and shape a backplate from Perspex or duralumin sheet about 90 mm square.

7 Apply the symbol with Tensol cement No. 6 to the backing piece.

ROAD SIGNS

TOILET SIGNS

OVER - SIMPLIFICATION ?

TRADEMARKS and SYMBOLS

BRITISH RAIL   BRITISH STEEL   BOAC "SPEEDBIRD"

 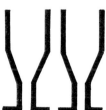 

WORLD CUP

AVON TOOLS   PETER DOMINIC - WINE MERCHANTS

FURTHER PUZZLES

Square with four identical pieces

Working on the principle that a puzzle should not be too easy to solve, design and make an assembly puzzle from thin plastic sheet based on the shape of a square. The puzzle must have four pieces of exactly the same size and shape which when assembled will form a square.

1 On paper draw a square with sides 80 mm long.
2 Divide the square into four parts of exactly the same size and shape, bearing in mind that the puzzle must not be too easy to solve.
3 Cut out the parts and apply them to Perspex with a suitable adhesive.
4 Roughly file the parts to shape.
5 Glue the pieces together one on top of the other.
6 With the four pieces together to ensure uniformity of size and shape, file the pieces to shape. The shaping must be done carefully as any fault will be repeated and could cause a bad fit.
7 Separate, remove all paper, clean the edges and polish.

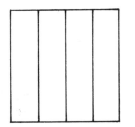 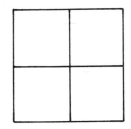

TOO SIMPLE

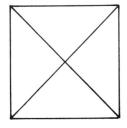 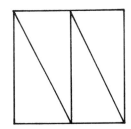

TOO SIMPLE

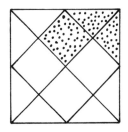 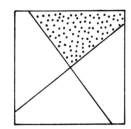

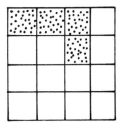 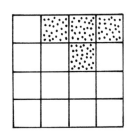

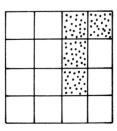 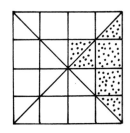

Triangle, rectangle and square

An isosceles triangle has an apex angle of 90°. Divide this triangle in such a way that the parts can be put together again to form both a rectangle and a square. Use this principle to make an assembly puzzle from plastic sheet.

1 On paper draw an isosceles triangle with an apex angle of 90°.
2 Find the middle point of each side. Join point *a* to point *b* and from *c* erect a vertical to meet *ab*.
3 Cut out the three pieces and fix them to some plastic sheet with adhesive.
4 File the pieces to shape and remove the paper. Finish and polish the edges.

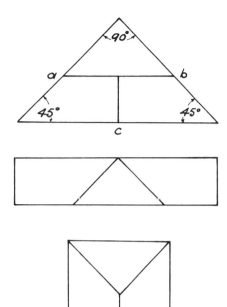

Another method

1 On paper draw an isosceles triangle with an apex angle of 90°.
2 Find the middle point of two sides and join points *a* and *c* and *b* and *c*.
3 Cut out the three pieces and fix them to some plastic sheet with adhesive.
4 File the pieces to shape and remove the paper. Finish and polish the edges.

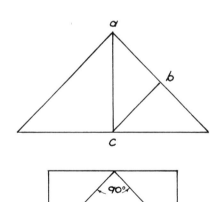

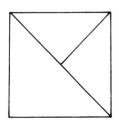

9

Square, L-shape and T-shape

Divide a square by drawing not more than three straight lines. The parts, when put together, must form either an L-shape or a T-shape with the same area as the square. Make a puzzle out of plastic sheet on this basis.

1 On paper draw a square with sides 80 mm long.
2 By drawing not more than three straight lines, divide the square so that the parts will form either a T-shape or an L-shape.
3 Cut out the pieces and fix them to plastic sheet with adhesive.
4 File the pieces to shape and remove the paper.
5 Finish the edges and polish.

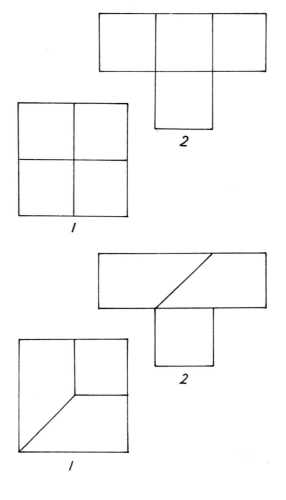

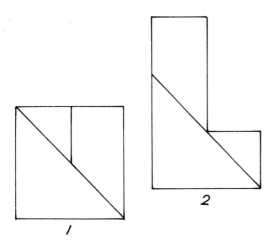

L-shape with four identical pieces

Draw an L-shape and divide it into four parts that are exactly the same in size and shape. With this information make a puzzle out of thin plastic sheet.

1 Draw a square with sides 80 mm long.
2 Find the middle point of each side and thus form an L-shape.
3 Divide the L-shape into four parts as shown in the drawing opposite.
4 Cut out these four pieces and glue them to thin plastic sheet.
5 Roughly file to shape.
6 Glue all the pieces together one on top of the other.
7 File the pieces to shape and size as one piece. This will ensure uniformity of shape.
8 Clean and polish the edges.
9 Separate the pieces, remove all paper and clean and round off all sharp edges.

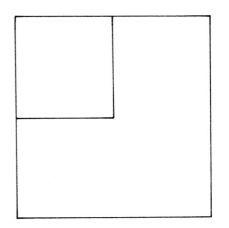

Three shapes in one hole

Draw a square with sides 40 mm long, a circle of 40 mm diameter and an equilateral triangle with a vertical height of 40 mm.

Suppose that holes exactly the same as these have been made in a piece of plastic sheet.

Can you make from a block of wood or plastic a shape that will pass through each of the holes filling each hole exactly as it passes through?

1   On a piece of thin plastic or metal sheet, draw a square with sides 40 mm long, a circle with a diameter of 40 mm and an isosceles triangle with a base of 40 mm and a vertical height of 40 mm.
2   Cut out these shapes using drills, junior saw and files.
3   Make a cylinder 40 mm diameter and 40 mm high using plastic rod, cast resin or wood dowelling.
4   Draw a diameter across one of the end faces.
5   Cut the cylinder to the shape shown in the drawing.

This finished shape will, in plan be a circle, in front elevation a triangle and in end elevation a square.

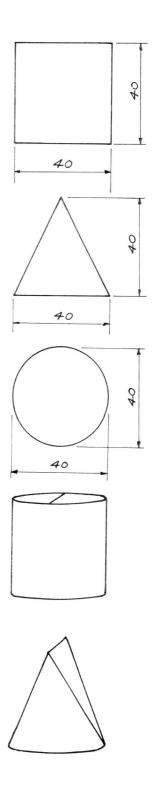

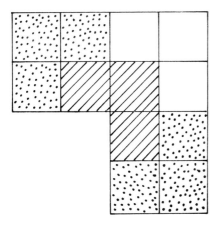

Divided tetrahedron

By dividing a tetrahedron (equilateral triangular pyramid) into two parts, make a two-part assembly puzzle out of cast resin or wood.

1 Make a tetrahedron with sides 40 mm long either by filing it out of wood or by casting resin into a mould.

The casting method

a Draw on thin cardboard a development of the three sides with a small lap for the joint.
b Fold up the shape and glue down the lap joint.
c Coat the inside with a resin sealer and then with a release agent. Cast in the resin pre-tinted to the required colour.

2 Mark out the lines shown on the drawing and cut along them with a fine saw.
3 File flat the two sawn surfaces and glue them together with paper interleaving.
4 True up the shape.
5 Separate the pieces, remove the paper and polish all surfaces.

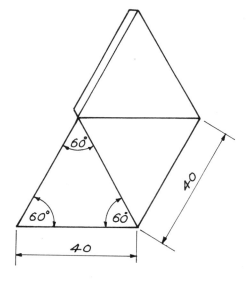

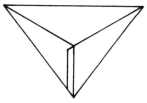

MOULD SOLDERED TOGETHER

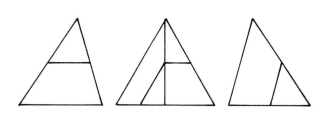

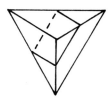

Four-part square

Design and make a four-part puzzle in plastic sheet that will assemble to form a square.

1 Draw, on paper, a square with sides 80 mm long.
2 Draw three or four straight lines within the square to form an interesting pattern.
3 Cut out the paper pieces and fix them to plastic sheeting with a suitable adhesive.
4 File the pieces to shape.
5 Trim the edges with wet and dry paper, remove the paper and polish with Perspex polish or Brasso.

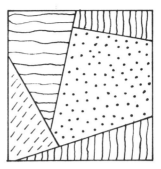

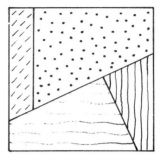

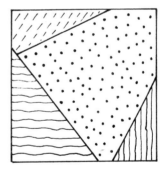

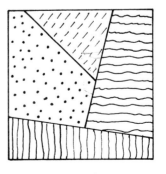

PROGRESSIVE DESIGN PROBLEMS

Mould for dice

Design and make a tinplate mould suitable for
making a dice out of cast resin.

1   Make a tinplate box by marking and cutting out
the development shown in the drawing.
2   Fold to shape using folding bars and soft solder
the seam.
3   Cut out the shape for the base and soft solder it
to the sides.
4   Mix the resin, tint as required, and make the
cast. Follow the manufacturer's instructions for
proportions of resin, pigment and catalyst.
5   When the resin has set, rip off the tin with a pair
of pliers.
6   File the dice to shape, rounding off all the corners. Finish the surfaces with wet and dry paper,
cutting compound and polish.
7   Mark out the position of the holes on pieces of
paper 25 mm square and fix these to the sides of the
dice with a suitable adhesive.
8   Drill the holes 3 or 4 mm diameter and 2 mm
deep and remove the paper.
9   Fill in the holes with cast resin tinted black.
10   Finish by polishing.

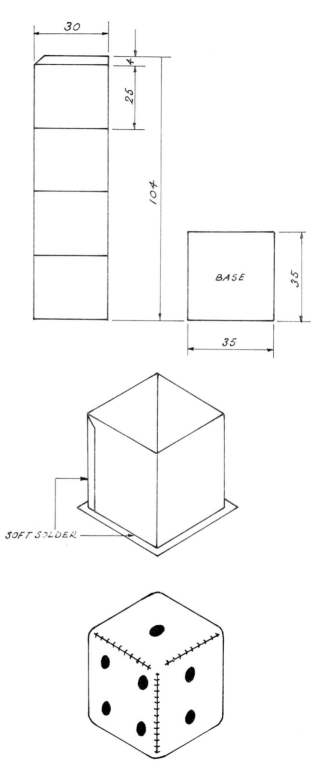

Ball-bearing toy

You have 4 or 5 ball bearings about 4 mm diameter with which you want to make a toy. Design and make a toy making use of these and some 3 mm plastic sheet.

1   On a piece of 3 mm thick Perspex sheet mark out the base 60 mm diameter and drill the holes for the ball bearings, forming a suitable pattern. The holes will need to be a little smaller than the diameter of the ball bearings.
2   Shape a piece of 3 mm or 1.5 mm clear Perspex sheet in a simple press, having first heated it to a temperature of about 150 °C in an oven or even with an open flame.
3   Trim and file the base edge of the curvature so that it lies flat on the base.
4   Trap the bearings inside and fix the two pieces together with Tensol cement No. 6.
5   File the waste off the base and polish.

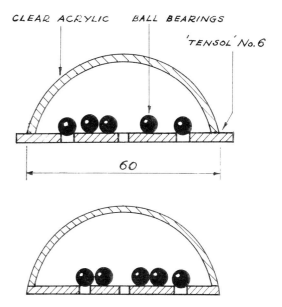

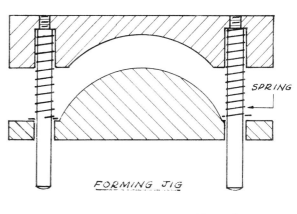

FORMING JIG

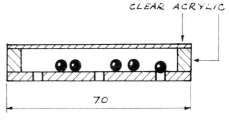

ALTERNATIVE DESIGN

Spinning top

Given a length of 4 mm diameter plastic or metal rod and a piece of plastic or metal sheet, make a spinning top that can be used in the same way as a dice.

1 Using bright mild steel sheet 1.5 mm thick or Perspex 3 mm thick, mark out a hexagon within a circle of 50 mm diameter.
2 Drill a hole in the centre 4 mm diameter to suit the rod.
3 Mark the numbers with punches if using steel or with an engraving tool if using plastic.
4 Cut and file to length the piece of plastic or metal rod. Make the length 40 mm.
5 Round off one end and taper the other using a file on the lathe.
6 Soft solder the metal rod to the plate making use of a soldering jig. If using Perspex, fix the rod with Tensol cement No. 6 or any other suitable adhesive.

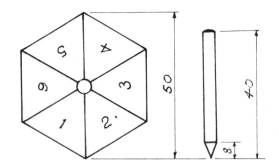

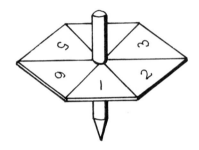

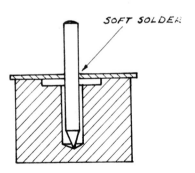

SOFT SOLDER

SOLDERING JIG

Teapot stand

Design a teapot stand in the form of a tile 150 mm square that can be made from cast resin.

1 Using a suitable wood, make a tray to the dimensions shown. Screw the sides to the base so that the sides can be taken off to release the cast.

2 a Seal the sides and base of the tray with a proprietry sealing compound.

or **b** Line the sides and base with Melinex plastic sheeting. Resin will not adhere to this plastic.

or **c** Line the base with Melinex and face the inner faces of the sides with a laminate such as Formica.

Faces not covered with Melinex will need coating with a release agent.

3 Mix a small quantity of gelcoat resin with a translucent pigment and catalyst and apply to the base of the tray.

4 When gelled, bed in two layers of 450 g/m² (1½ oz) mat using a standard lay-up resin.

5 On this base, lay out a decoration, e.g. photograph, painting, leaves, glass chippings, pieces of Perspex, etc.

6 Calculate the approximate volume of casting resin necessary to fill the mould. Mix this according to instructions, tint with translucent pigment if desired and cast to fill the mould. Place a film of Melinex over the surface to exclude the air, otherwise the surface exposed to air remains tacky when the cast has set.

7 When set, extract from the moulding box, trim and polish.

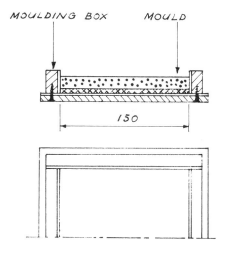

MOULDING BOX   MOULD

150

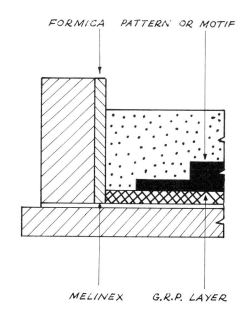

FORMICA   PATTERN OR MOTIF

MELINEX   G.R.P. LAYER

ENLARGED SECTION

17

2 Pendant and ring design

PENDANTS

Pendant from Perspex

From a piece of Perspex 60 mm×40 mm×3 or 5 mm thick, make a pendant.

1 File the Perspex to a size of 60 mm×40 mm and round off all sharp corners and edges.
2 The decoration decided on must be evenly balanced so that the pendant will hang straight when it is suspended. Use Tensol cement No. 6 to fix any superimposed parts.
3 Finish the surfaces with wet and dry paper and polish with Perspex polish.
4 Make a suspension ring by winding 1 mm diameter wire (brass or stainless steel) around a suitable former of 5 mm diameter.
5 When mounting the ring on the pendant, leave it twisted out of alignment, fix it on and then twist the two ends together with a pair of pliers. Never pull the two ends apart and consequently spoil the roundness of the ring.

DRILLED HOLES

DRILLED HOLES
FILLED WITH WOOD,
METAL OR PLASTIC
FILED FLUSH

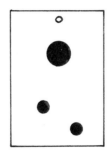

DRILLED HOLES
FILLED WITH WOOD,
METAL OR PLASTIC
LEFT PROTRUDING

MOTIF SUPER-
IMPOSED

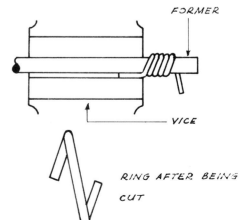

FORMER

VICE

RING AFTER BEING
CUT

Pendant from metal

Make a pendant from a piece of brass, copper or dural sheet 60 mm×45 mm×1 mm thick.

1 File the metal to a size of 60 mm×45 mm with all the edges straight and square to each other.
2 Draw two or three straight lines to break the rectangle into an interesting shape. Make all the lines of different lengths to create interest.
3 Cut off the waste and file to shape, rounding off all sharp edges and corners.
4 Decide where the suspension hole is to be and drill this hole 3 mm diameter.
5 The design can be made more interesting by hammering the surface of the metal, drilling holes or applying pieces of metal or plastic.
6 The suspension ring can be made as in the previous example.

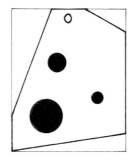

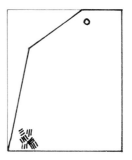

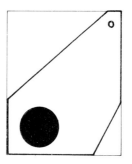

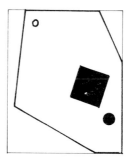

IDEAS FOR PENDANTS

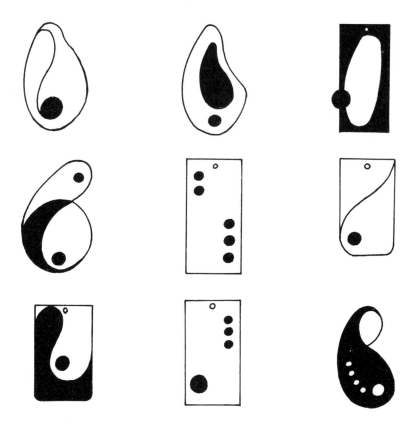

PENDANTS MADE OF METAL SHEET USING
APPLIED STRIP, PLANISHING MARKS AND
DRILLED HOLES FOR DECORATION

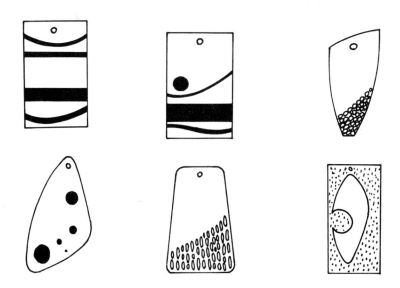

METHODS OF FIXING NATURAL STONES

Araldite or jeweller's cement

Ensure that the surfaces are clean and roughen the base of the stone and the face of the metal to provide a key.

Claws A

1 Scribe round the stone and mark 3, 4, or 5 claws 2 or 3 mm wide on the baseplate.
2 Shape the plate with a piercing saw, tension file etc.
3 File the edges and polish.
4 Bend each claw at right angles to fit the stone.
5 Close down the claws with a punch or a pair of pliers.

Claws B

1 Mark round the stone onto the metal, holding the stone in position with a spot of Plasticine or Twinstick.
2 Drill 3, 4 or 5 holes 1.5 mm diameter with the circumferences of the holes touching the outline.
3 Make the claws from suitable wire and taper them to a push fit in the holes.
4 Arrange the claws so that they all project the same distance and soft solder them in position from the back.
5 File the claws flush at the back.
6 Taper and round off the end of the claws and close them down onto the stone.

Bezel fitting

1 Make a narrow ring from thin metal. This ring must be the correct size to fit the stone. Solder the ring joint with hard running silver solder.
2 Solder the ring to the baseplate with Easy Flo' silver solder or soft solder.
3 Position the stone and burnish the edge of the ring over the stone.

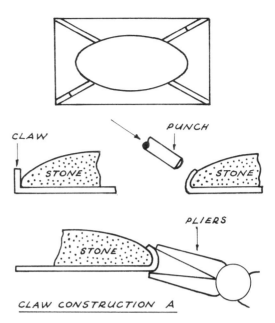

CLAW CONSTRUCTION A

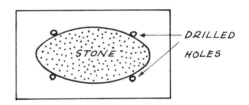

CLAW CONSTRUCTION B

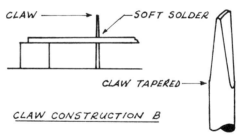

BUTT JOINT SOLDERED WITH
HARD RUNNING SOLDER

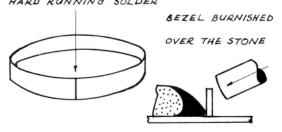

BEZEL FITTING

BROOCHES

Fish brooch

From a piece of copper or nickel silver 1 mm thick make a brooch in the shape of a fish that would be suitable for your mother or sister.

1 Draw the shape of the fish on paper.
2 Fix the paper pattern to the metal with suitable adhesive.
3 Drill three or four small holes to take the claws that will hold the stone in position.
4 Drill any other holes that form part of the decoration.
For designs 1 and 2 drill holes part way through the metal. These holes will be filled later with black wax crayon or tinted resin.
5 File to the shape of the pattern and soft solder the claws in position.
6 Soft solder the brooch pin to the back or glue the pin to the back with Araldite.
7 Fill any holes with resin or crayon.
8 Taper the claws and fit the stone.
9 Polish the surfaces and apply a thin coat of clear lacquer if copper has been used for the backplate.

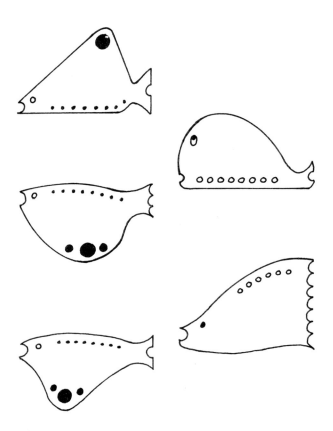

22

Brooch with natural stone

Make a small brooch incorporating a natural stone in a claw setting.

1   Mark out the pattern on paper.
2   Fix it to the metal with suitable adhesive.
3   Drill three or four small holes to suit the size of the claw wires, positioned near the periphery of the stone.
4   File to the shape of the pattern.
5   Cut the radial lines with a pair of snips or with a junior hacksaw.
6   Soft solder the claws in position.
7   File and clean off the waste of the claws from the back and taper the projecting claws.
8   Solder the brooch pin to the back or glue the pin with Araldite.
9   With a pair of pliers bend the segments at slightly different angles.
10   Set in the stone by bending the claws.
11   Polish and lacquer if necessary.

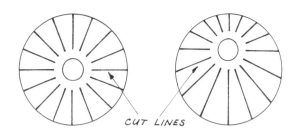

CUT LINES

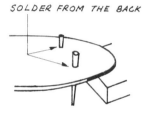

SEGMENTS BENT AT
DIFFERENT ANGLES

SOLDER FROM THE BACK

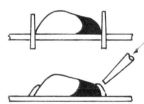

FITTING THE CLAWS

FIXING THE STONE

23

DRESS RING

Make a dress ring from 18 s.w.g. or 1 mm thick nickel silver, fitted with one natural gem stone.

1   Make the ring from a piece of tube of the right diameter.

or

2   Bend the ring from a piece of strip metal. To find the length of metal to make the ring, wrap a strip of paper around the finger and cut the metal 3 mm less than this length. Shape the ends and bend the strip around a piece of rod of the correct diameter.

or

3   Find the length of metal required as explained above. File the ends square and bend the strip around a piece of rod of the required diameter. Silver solder the joint.

4   Clean and finish the surfaces and edges.

5   Fix the stone:

    a with adhesive – Araldite or jeweller's cement.

or **b** with adhesive to a baseplate the same shape as the stone.

or **c** with adhesive to a baseplate a different shape from the stone.

or **d** by means of claws shaped in one with the baseplate.

6   Fix the baseplate to the ring with soft solder or silver solder. Wire the plate to the ring with fine binding wire for soldering and trap a small pellet of solder between the plate and the ring.

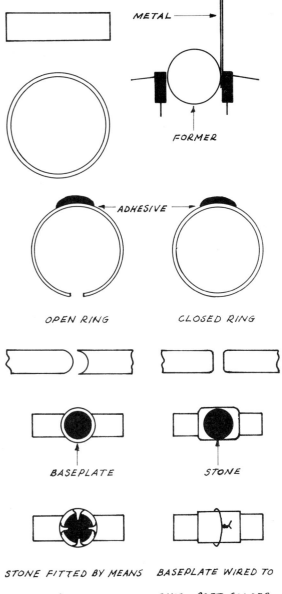

METAL

FORMER

ADHESIVE

OPEN RING   *CLOSED RING*

BASEPLATE   *STONE*

STONE FITTED BY MEANS OF CLAWS

BASEPLATE WIRED TO RING. SOFT SOLDER OR SILVER SOLDER JOINT

3 Design problems in functional and free-standing items

CANDLE HOLDER

Design and make a fitting to hold one or two candles of 12 mm diameter. The only materials you may use are BDMS rod 5 mm and 14 mm diameter.

1 Face the ends of the 14 mm rod in the lathe and make the length 25 mm.
2 Drill one end 12 mm dia. and 18 mm deep.
3 Drill the other end 4 mm dia. and tap 5 mm.
4 Some of the shapes are best machined by making a chucking allowance and then parting off when the machining is completed.
5 Anneal the 5 mm rod and bend it to shape around suitable formers.
6 Thread the top end 5 mm.
7 Screw the holder to the base and soft solder the pieces together or 'lock' the thread with Araldite.

A holder for two candles

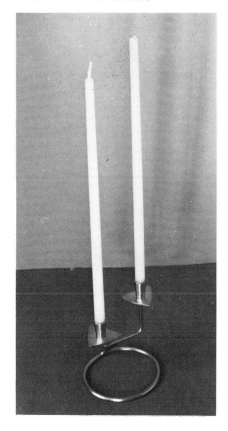

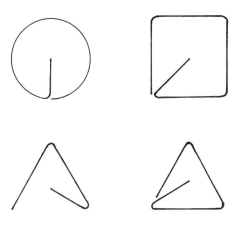

BASE SHAPES

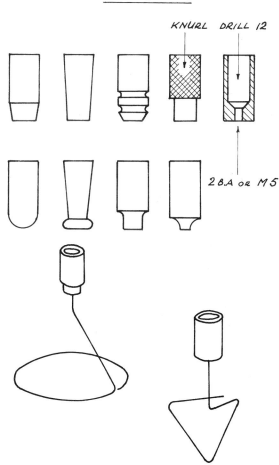

KNURL DRILL 12

2 B.A or M 5

25

EGG STAND

From a piece of duralumin sheet 18 mm thick make a stand to hold a boiled egg.

1 Mark out the shape having first decided on the size of the hole to take the egg.

2 Cut the hole with a piercing saw, tension file, hole saw or by drilling a series of small holes and then filing.

3 File the outer shape.

4 Polish the surfaces and edges while the metal is still in the flat.

5 Bend the shape carefully using folding bars to avoid marking the metal or bend around a suitably shaped hardwood pattern.

A stand to support a set of the egg-holders can be made from wood or a combination of wood and plastic.

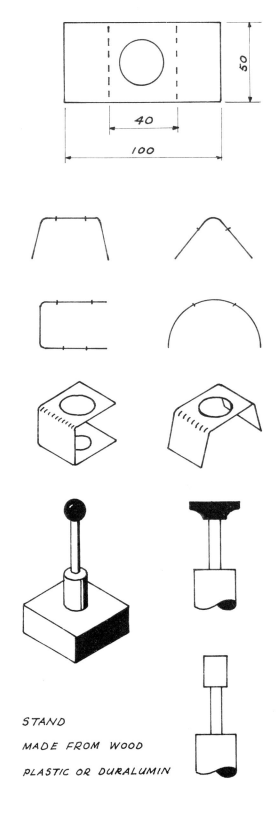

STAND

MADE FROM WOOD

PLASTIC OR DURALUMIN

EGG TIMER HOLDER

Design a fitting to hold an egg timer using either of the following principles:

a resetting the timer by rotating it in the holder, *or*
b resetting the timer by turning the complete fitting.

a Use acrylic sheet or duralumin 3 mm thick and a Terry clip for holding the sand glass.
Slot the feet into the upright and fix the joint with Tensol cement No. 6.

b Use acrylic sheet 3 mm thick and after heating to about 150 °C bend to shape around a suitable former.
Fix the sand glass by seating it in two washers made of Perspex or fibre about 3 mm thick.

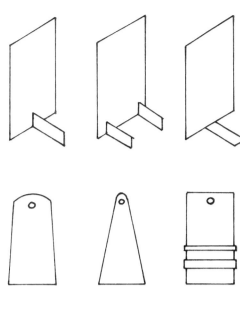

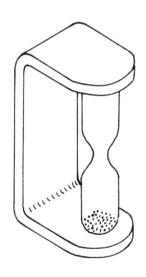

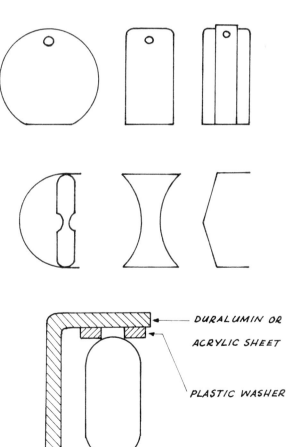

DURALUMIN OR
ACRYLIC SHEET

PLASTIC WASHER

COCKTAIL STICKS AND STAND

Design and make a set of four cocktail sticks and a stand to hold them.

1 Use plastic 6 or 8 mm thick. On this mark out the handle shapes.
2 Drill blind holes the correct size to take the sticks.
3 File the handles to shape and finish with wet and dry paper and then polish.
4 For the sticks use wooden cocktail sticks or make them from thin plastic rod or small plastic knitting needles about 3 mm diameter.
File the rods to a taper on the lathe and finish with wet and dry.
5 Fix the sticks to the handles with Tensol No. 6 cement.

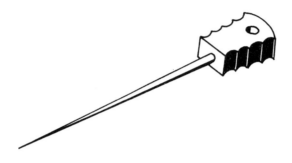

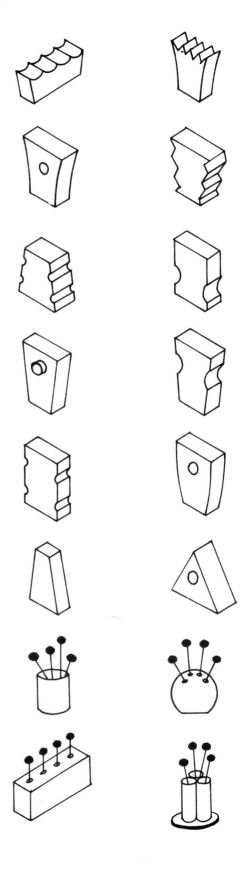

Suggestions for stands

1 Plastic pill boxes.
2 Wood, metal or plastic to a choice of shape.
3 Cast resin – a wide range of shapes and colours possible.
4 Clear cast resin with encapsulated motifs or flowers.
5 Drill 4 holes in the stand an easy fit for the sticks.

SIMPLE CANDLE HOLDER

Make a small candle holder from a piece of BDMS sheet 1 mm thick and a length of steel tube 16 mm diameter.

1 Either mark out the shape on the BDMS sheet or draw it on paper first and fix it to the sheet with adhesive.
2 Cut the sheet to shape and finish the surfaces and edges.
3 At this stage make any bends that are necessary using a former of the correct diameter.
4 File the ends of the tube or face the ends in the lathe.
5 Round off any sharp edges and polish.
6 Soft solder the tube to the base.
7 Wash off the flux and clean the surfaces.
8 Finish with a clear or coloured lacquer or paint with an egg-shell finish paint.

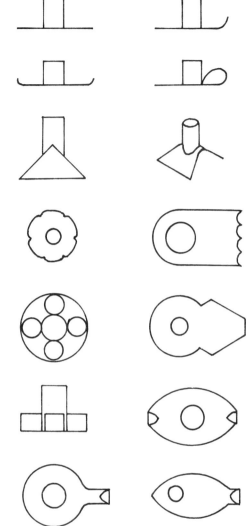

A simple candle holder

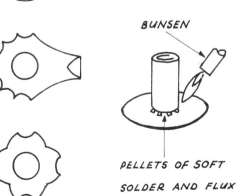

BUNSEN

PELLETS OF SOFT SOLDER AND FLUX

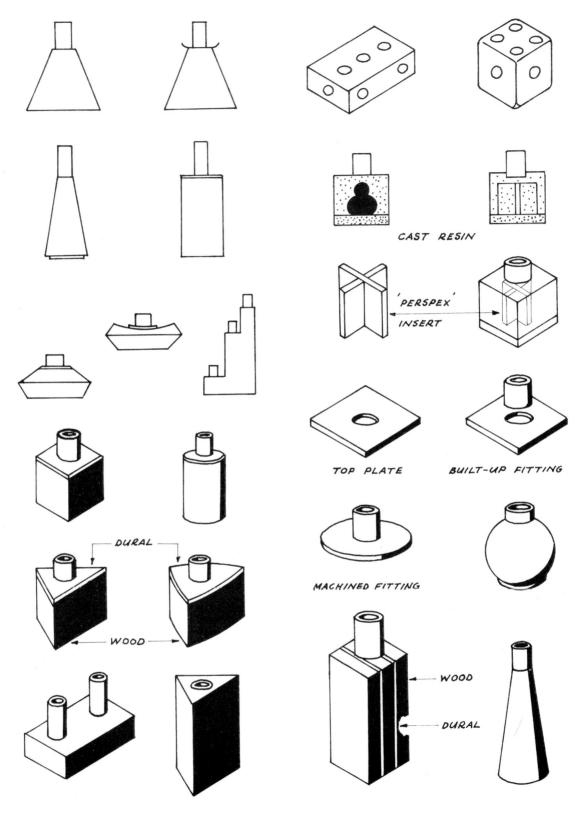

CAST RESIN

'PERSPEX' INSERT

TOP PLATE

BUILT-UP FITTING

MACHINED FITTING

DURAL

WOOD

WOOD

DURAL

CANDLE HOLDER

Using as a base a block of hardwood or cast resin make a fitment to hold one or two candles of 16 mm diameter.

Wood base

Machine, plane or chisel the base according to the final shape decided on. If the candles are to be fitted directly into the wood, drill the holes the correct size to fit the candles using a depth stop if more than one candle is being used.

Metal tube and (or) metal sheet, e.g. duralumin, may be used to protect the wood face. The tube and sheet can be machined in one piece from the solid or built up and the joint made with Araldite. Use Araldite too as an adhesive to fix the metal to the wood.

Cast resin base

1 Make a mould to the size and shape required from tinplate (see under 'Mould for dice', page 14) or use a suitably shaped porcelain or polyethylene container. Remember that a square prism of resin has better reflective qualities than a cylinder.

2 Coat the inside surfaces of the container with a release wax.

3 Measure out a quantity of resin sufficient to form a base layer. Add translucent or opaque pigment and the correct amount of catalyst. Stir well and wait for any bubbles to rise before pouring it into the mould.

4 When the base layer has set, insert the motif that is to be encapsulated. Translucent Perspex, built into a pattern, is very effective for this purpose. Remove any sharp edges and corners because these encourage cracks to develop. Pour in resin to the depth required. If the cast is large in volume, considerable exothermic heat is generated which will cause cracks to develop in the resin as it cools. The addition of too much catalyst will aggravate this. Partially immersing the mould in a cold water bath will help to dissipate the heat.

IDEAS FOR CANDLE HOLDERS

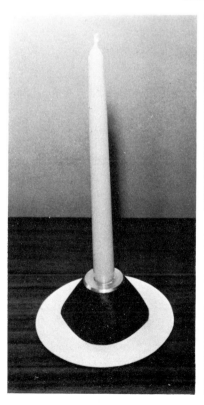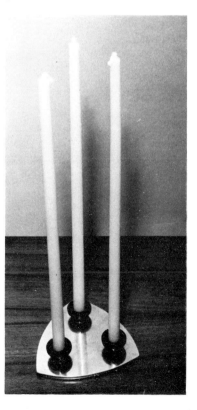

IDEAS FOR CANDLE HOLDERS

TABLE NAPKINS

Design and make a holder or holders suitable for table napkins.

MATERIALS – metal tube, metal sheet, wood, Perspex, cast resin.

Round metal tube. Split the tube or leave it solid. Machine the tube to a suitable shape on the lathe. Decorate by chasing, engraving, punching, drilling, etching or filing.

Square or rectangular tube. Split the tube or leave it solid. Very effective when left plain and painted in bright colours with either a matt or satin finish paint.

Metal sheet. Bend to any suitable shape around wood or metal formers. Leave the ends open or bring them together and silver solder or braze the joint.

Wood. Use a close-grained hardwood with a good natural colour. Shape by hand methods or turn to shape on a lathe.

Cast resin. A container is needed into which to cast the resin. Make this out of tinplate with soft soldered joints. A length of tube or rod fixed to the base can act as a core to form the centre hole.

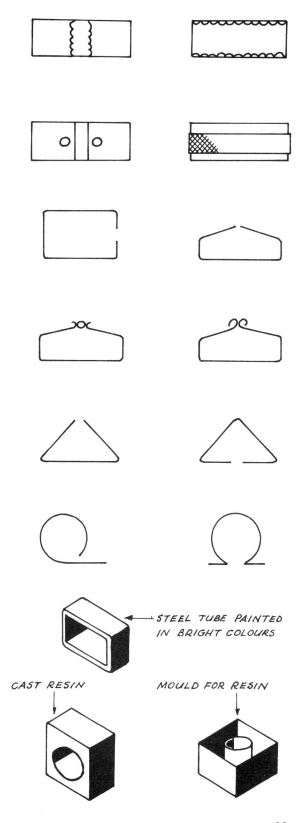

STEEL TUBE PAINTED IN BRIGHT COLOURS

CAST RESIN

MOULD FOR RESIN

FRYING PAN SLICE

Design and make a frying pan slice, developing the idea in relation to the shape of a frying pan.

1   A straight stem and blade, with the blade profile having no relation to the outline of the pan.
2   The blade bent so that the blade can lie flat in the pan and the blade shaped in relation to the pan.
3   Blade left straight, stem bent.
4   A double bend in the stem.

The pattern formed by the draining holes should have some relation to the outer shape.

BLADE – Aluminium alloy sheet 1 mm thick.
STEM – Aluminium alloy rod 6 mm dia. or aluminium alloy strip 12 mm×3 mm.
HANDLE – Hardwood or Erinoid or cellulose acetate rod 16 mm diameter.

To provide drainage it is easier to drill holes than to form slots.
Fix the stem to the blade by means of two snaphead rivets 3 mm diameter.
A stem made of rod will give more rigidity than one made of strip, but the end will need to be forged out to take the two rivets.
An epoxy adhesive applied to the joint before riveting will ensure that the joint remains firm during use.
Clean the alloy surfaces with a fine grade of steel wool.

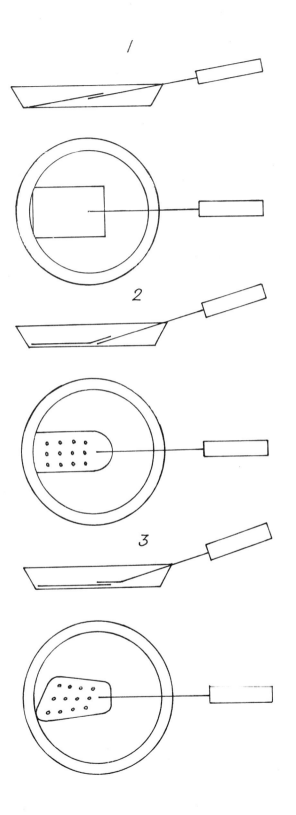

Fixing the handles

1 Plastic handles

File a square tang on the end of the stem making the end about 4 mm square. Drill a hole in the end of the rod 5.5 mm diameter and slightly deeper than the length of the tang.

Shape the plastic on the lathe and polish with pumice powder and oil or a proprietary cutting compound. Soak the plastic in boiling water to soften it and push it steadily onto the tang. Check for alignment and repeat until the tang enters fully into the handle. Cool off under running cold water.

2 Hardwood handles

Machine grooves in the end of the stem or knurl the end or groove and knurl. This will provide a key for the adhesive.

Shape the handle and drill a hole in the end slightly deeper than the length of the tang and of a size that will allow the handle to be a push fit. Fix the handle to the stem with an epoxy adhesive.

Alternative fixing. File a tang on the stem as shown. Drill a hole in the handle and file a groove in it so that it fits the tang. Fix the handle by means of rivets and an epoxy adhesive.

If a lathe is not available to shape the handle, use strip metal for the stem and fix strips of wood or plastic to it with rivets and an epoxy adhesive.

Finish the wooden handles with either linseed oil or a polyurethane lacquer.

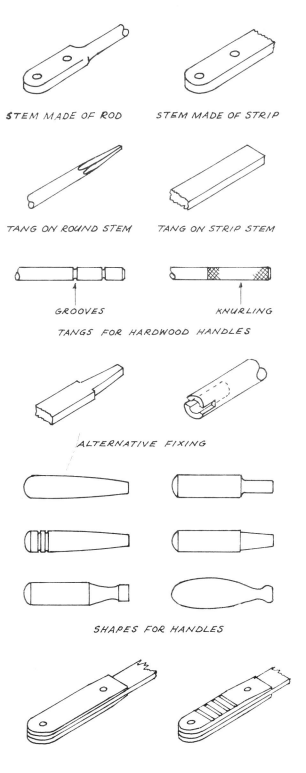

STEM MADE OF ROD   STEM MADE OF STRIP

TANG ON ROUND STEM   TANG ON STRIP STEM

GROOVES   KNURLING

TANGS FOR HARDWOOD HANDLES

ALTERNATIVE FIXING

SHAPES FOR HANDLES

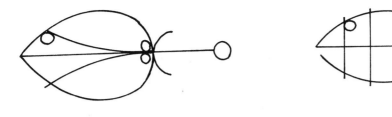
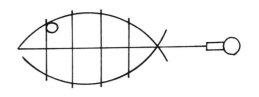

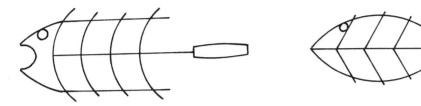
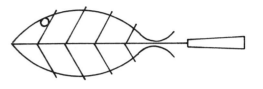

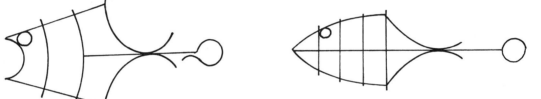
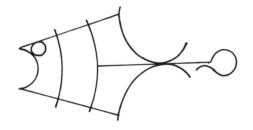

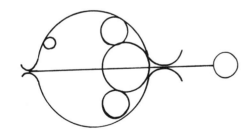
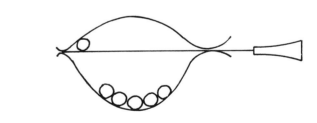

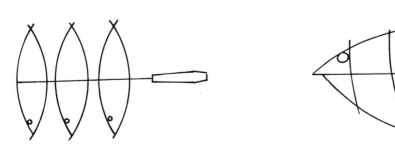
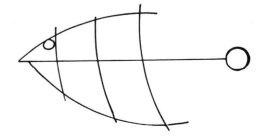

36

TRIVET

Design a trivet suitable for holding hot fish dishes using 5 mm diameter mild steel rod.

Keep the shape simple and symbolic.
Bend the rod around suitably shaped formers making sure that all joints come together tightly.
Parts of the framing can be extended to form legs and these can terminate in plastic feet.
All the joints may be brazed or key joints brazed and the remainder soft soldered.
Joints must be clean and must fit well. They can be held in position for soldering by means of soft iron binding wire.
Form the eye from a short length of tube soft soldered or brazed in position.

Finish. Heat resistant enamel, Hammerite paint or polyurethane paint.

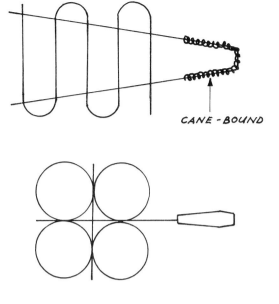

CANE - BOUND

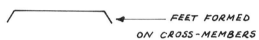

FEET FORMED
ON CROSS-MEMBERS

Handles

A low conductivity material is best for the handle but metal can be used. It can then be insulated by binding with cane, plastic strip or cord or by plasticoting.

Securing the handle

1 *Wooden handle.* Use an epoxy adhesive.
2 *Plastic handle.* Drill a hole in the handle 0.5 mm smaller than the diameter of the stem. Expand the plastic by soaking it in boiling water for a few minutes and pushing it onto the stem. Cool off under running cold water. Alternatively, use a suitable plastic adhesive.
3 *Metal handle.*
    a Cut an internal thread and screw the handle on,
or **b** use an epoxy adhesive,
or **c** soft solder or braze the joint.

HANDLES MAY BE SHAPED FROM METAL, PLASTIC OR WOOD, PLASTICOTED OR BOUND WITH CANE, CORD OR PLASTIC

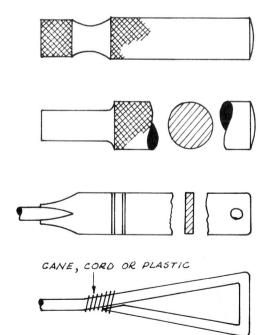

CANE, CORD OR PLASTIC

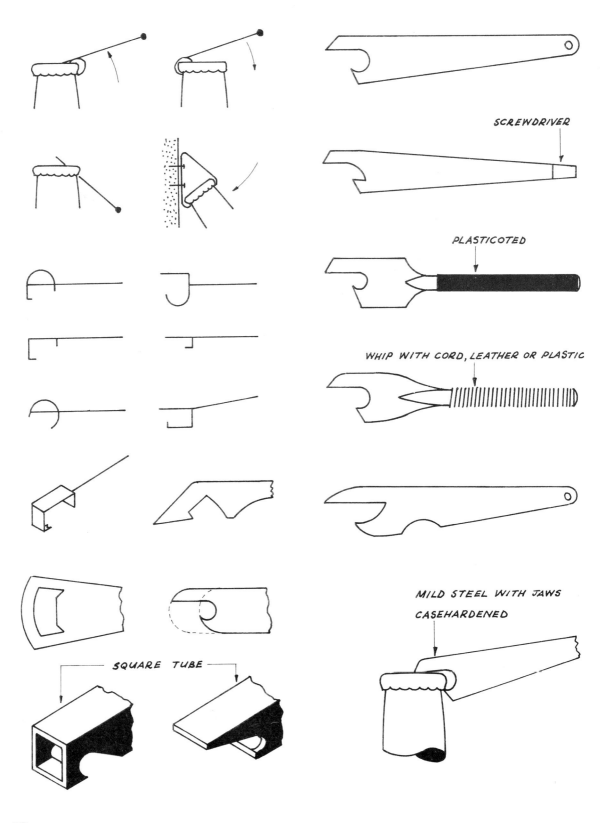

SCREWDRIVER

PLASTICOTED

WHIP WITH CORD, LEATHER OR PLASTIC

MILD STEEL WITH JAWS
CASEHARDENED

SQUARE TUBE

BOTTLE OPENER

Design and make a tool that will take the metal cap off a bottle.

The tool can be designed to operate by levering upwards or downwards.
Remember that a tool incorporating internal shaping is more difficult to make than one with external shaping.

MATERIAL – Steel or duralumin. Case-hardening the operating end of an opener made of mild steel will increase its working life.

Jointing

Soft solder, braze or rivet. Remember that duralumin cannot be satisfactorily soft soldered and can be hard soldered only with difficulty. A handle can be joined to a blade by slotting the handle, shaping the end and brazing the joint.

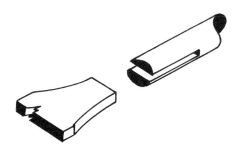

Handles

Finish can be:

a natural steel,
b by plasticoting,
c by binding with cord, leather or plastic thonging,
d painting.

CHESS SET

Design a set of chess-men and a board on which to use them.

MATERIALS – Wood, plastic rod, cast resin, metal rod, tube or plate.

Because of the need to produce a number of components exactly the same, keep the shapes simple. This does not apply if flexible moulds are used.

Design 1

BASE – Plastic, wood or metal.
CHESS MOTIFS – Plastic, duralumin, nickel silver or brass.

Machine the base pieces in one length and part off the required parts.
Cut and file the motifs to shape using templates to ensure uniformity of shape. Form the slots either on a shaping machine or on a milling machine or by hand methods. If hand methods are used, make a jig so that all the slots are uniform.
Fix the motif to the base by means of an epoxy resin.

Design 2

MATERIALS – Duralumin or duralumin and plastic.

Keep the shapes simple and keep to shapes that are easy to repeat by machining methods.
Templates may be necessary to ensure uniformity.
Knurl the rings in one piece with the main body, or machine the knurled rings separately and shrink or glue them in position.
File the motifs to shape.

1

METAL PLATE GROOVED INTO A PLASTIC OR WOODEN BASE

2

DURALUMIN OR PLASTIC ROD

KNURLED IN THE SOLID OR A KNURLED SLEEVE SHRUNK ON

Design 3

MATERIALS – Plastic rod and duralumin tube.

Make templates to ensure uniformity of shape.
Fit the duralumin tube by shrinking it on or fix with an epoxy adhesive.
Form the motifs by machining or by filing them to shape.
A method must be evolved to ensure that the lengths and diameters that are machined on all pieces are uniform.

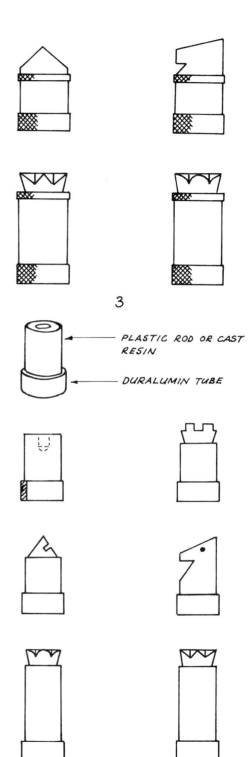

3

PLASTIC ROD OR CAST RESIN

DURALUMIN TUBE

Design 4

MATERIALS – Hardwood or plastic and duralumin plate.

Use an epoxy adhesive to fix the layers of material together.

Assemble the sheets of material and then cut out the individual shapes.

Note. Flexible moulds made of either Silicone rubber or latex permit the easy casting in plaster or cast resin of complicated shapes. Metallic fillers used with casting resin offer a wide scope.

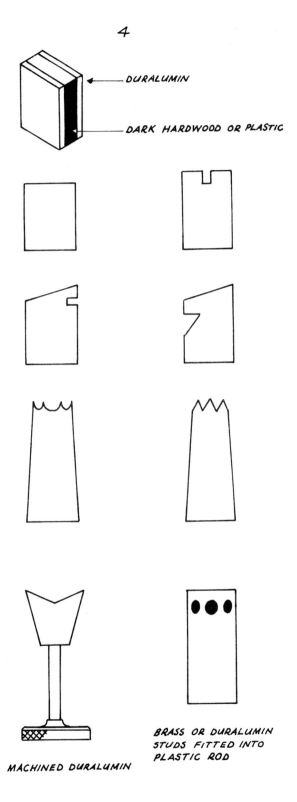

4

DURALUMIN

DARK HARDWOOD OR PLASTIC

MACHINED DURALUMIN

BRASS OR DURALUMIN STUDS FITTED INTO PLASTIC ROD

BASEBOARDS

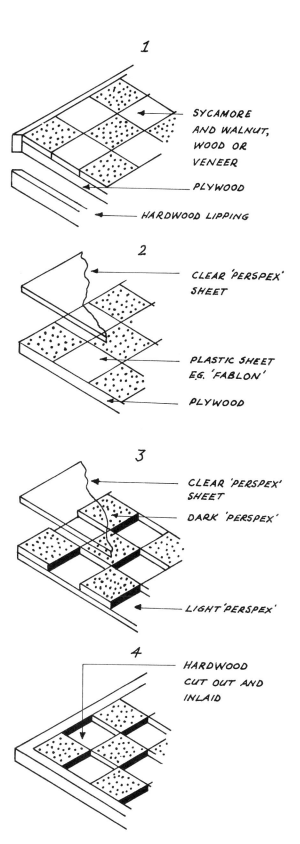

Example 1

Use plywood for the base, the thickness being governed by the size of board required. Contrasting coloured hardwoods 4 or 5 mm thick or contrasting veneers, e.g. sycamore and walnut, form the squares. The square shapes must be cut accurately to ensure uniformity of size. Consider designing a jig for this. Using a suitable adhesive, fix hardwood lipping around the edges.

1

SYCAMORE AND WALNUT, WOOD OR VENEER

PLYWOOD

HARDWOOD LIPPING

Example 2

This is a quick, easy method of construction. Use plywood or hardboard for the base. Form the squares from self-adhesive plastic sheet in contrasting colours. Stick the squares to the base and cover the whole surface with clear Perspex sheet 2 or 3 mm thick. Lipping will be needed around the edges.

2

CLEAR 'PERSPEX' SHEET

PLASTIC SHEET E.G. 'FABLON'

PLYWOOD

Example 3

Use a light-coloured opaque 'Perspex' sheet 3 or 4 mm thick for the base. Cut the dark squares from opaque Perspex sheet 3 mm thick. Glue these in position with Tensol cement No. 6, leaving the colour of the base piece to form the lighter coloured squares. Face all the surface area with clear Perspex sheet 2 or 3 mm thick.

3

CLEAR 'PERSPEX' SHEET

DARK 'PERSPEX'

LIGHT 'PERSPEX'

Example 4

This is made from a solid piece of good quality hardwood. Mark out the squares on the hardwood and chisel out alternate squares to an even depth. Inlay a contrasting wood or veneer into these recesses. Scrape the top surface to a suitable finish. Edge lipping will be necessary.

4

HARDWOOD CUT OUT AND INLAID

SCREWDRIVERS 1

BLADE – High carbon steel or silver steel 5 mm diameter.

HANDLE – Erinoid or cellulose acetate rod (fluted or hexagonal, if possible) 16 mm diameter.

File the blade to shape making the end 0.5 mm thick. Keep the faces flat. Emery cloth the surfaces to get rid of all scratches and file marks. Heat the end to a bright red using a gas-air torch and quench in water. Brighten the surfaces with emery cloth and, using a bunsen burner, temper the end to a brown colour. Quench in water when the correct colour reaches the tapers.

Machine the handle to a suitable shape using a lathe and drill the hole for the blade a little deeper than the length of the tang with a 4.5 mm drill. Soak the handle in boiling water and push the blade home. Check for straightness and repeat the process if the tang has not entered fully into the handle. Cool off under running cold water.

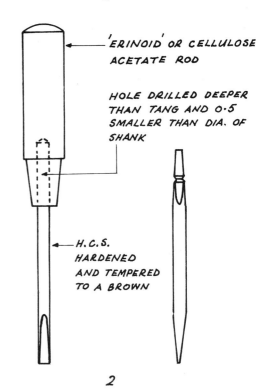

1

'ERINOID' OR CELLULOSE ACETATE ROD

HOLE DRILLED DEEPER THAN TANG AND 0·5 SMALLER THAN DIA. OF SHANK

H.C.S. HARDENED AND TEMPERED TO A BROWN

SCREWDRIVERS 2

BLADE – High carbon steel or silver steel 6 mm diameter.

HANDLE – Erinoid or cellulose acetate hex. 16 a/f.

COLLAR – Brass rod 16 mm diameter.

Forge the blade to shape, hammering the metal only when it is red-hot. Anneal the metal by heating to a bright red and cooling off slowly under hot firebricks or hot ashes. File the blade to shape. Finish the surfaces with emery cloth and harden and temper as above.

Machine the collar to shape and silver solder it to the blade.

Machine the handle and fix it to the blade in the same way as Screwdriver 1. Cross rivet the handle to the blade with a 1.5 mm diameter steel rivet.

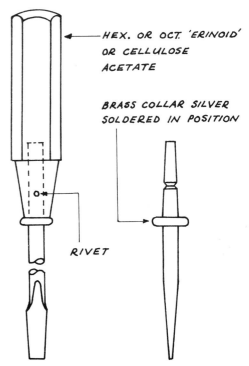

2

HEX. OR OCT. 'ERINOID' OR CELLULOSE ACETATE

BRASS COLLAR SILVER SOLDERED IN POSITION

RIVET

SCREWDRIVERS 3

BLADE – High carbon steel rod 6 mm diameter.
HANDLE – BDMS hex. 16 mm a/f.

Forge, shape and harden and temper the blade in the same way as Screwdriver 2.
Machine the handle to shape on a lathe.
Drill out the centre as large as possible to lighten the weight.
Drill the handle to take the blade using a 5.9 drill so that the blade will be a force fit in the handle.
Turn a plug so that it is a press fit in the top end of the handle and shape the end with a form tool or use a file and emery cloth.
Rivet the handle to the blade with a 1.5 mm diameter rivet.

SCREWDRIVERS 4

BLADE – High carbon steel rod 6 mm diameter.
HANDLE – Duralumin or aluminium rod 16 or 18 mm diameter.

Forge, shape and harden and temper the blade in the same way as Screwdriver 2. Knurl a short length of the blade near the handle end.
Knurl and shape the handle on a lathe.
Drill it so that the blade is a force fit in it. Rivet the handle to the blade with a 1.5 mm diameter steel rivet.

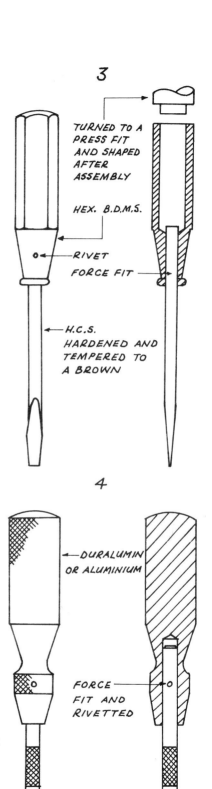

SCREWDRIVERS 5

BLADES – High carbon steel rod 6 mm diameter.
HANDLE – Aluminium or aluminium alloy rod 18
    mm diameter.
COLLET NUT – Aluminium alloy or brass rod 16 mm
    diameter.

Shape and harden and temper the blade. Cut away
the tang of the blade to half its diameter.

Knurl and taper the handle and turn the collet with
a 60° included angle taper. Screw cut it with a 12
mm diameter fine thread.

Cut the slots to form the collet on a milling
machine or saw them and finish them with a ward-
ing file.

Machine the collet nut, using a centre drill to
obtain the 60° taper. Tap the hole and knurl the
outer surface.

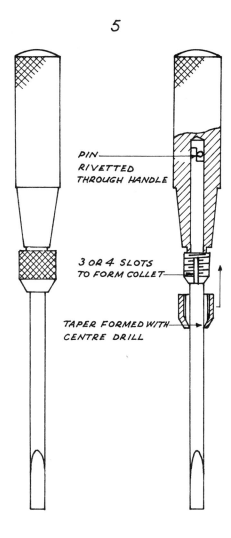

5

PIN RIVETTED THROUGH HANDLE

3 OR 4 SLOTS TO FORM COLLET

TAPER FORMED WITH CENTRE DRILL

SCREWDRIVERS 6

BLADE – High carbon steel rod 6 mm diameter.
HANDLE – Aluminium or aluminium alloy rod 18
    mm diameter.
LOCKNUT – Aluminium or aluminium alloy rod 16
    mm diameter.

Make the blade in the same way as the previous
examples. The lugs that project from the blade and
prevent its rotation present a constructional problem.

Drill and ream a hole in the handle so that the
blade is a sliding fit in it. Thread the end of the
handle with a 12 mm diameter thread and cut a slot
across it.

Tap the inside of the locknut to fit the handle and
knurl the surface.

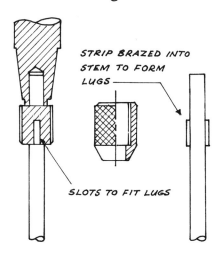

6

STRIP BRAZED INTO STEM TO FORM LUGS

SLOTS TO FIT LUGS

4 Progressive design

SCREWDRIVERS 7 – TWO-PIECE MOULD

BLADE – High carbon steel rod 4 or 5 mm diameter.
HANDLE – Cast resin.

Shape the blade and harden and temper it.
Make the moulding box as shown in the drawing and then saw it in half to form a top and bottom. Fit two locating dowels to line up the two halves.
Obtain or make a screwdriver with a blade of 4 or 6 mm diameter. Coat the surfaces with a wax release agent and position it in one half of the moulding box. Fill the box with a cast resin and, when set, make the top surface flat.
Repeat with the other half of the box.
Withdraw the pattern, fix the two box halves together and drill a pouring hole.
Cramp together the two parts of the mould, putting wax release agent on the mating surfaces and with the blade positioned between them. Fill with resin and allow to set.

SCREWDRIVERS 8 – TWO-PIECE MOULD

BLADE – High carbon steel 5 mm diameter.
HANDLE – Cast resin.

Make the mould from two blocks of steel, aluminium or aluminium alloy.
Fix the two parts together with six 6 mm diameter nuts and bolts, using two dowel pins for location.
Internally shape the handle by drilling the small holes first and then the larger one.
Coat the internal surfaces of the mould with wax release agent and fix the two parts together with the blade in position. Cast the resin with the blade in position.
Extract from the mould and finish shaping the end of the handle on a lathe.

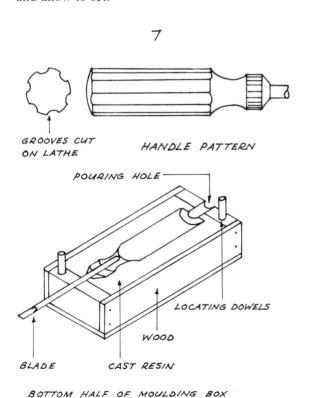

7

GROOVES CUT ON LATHE   HANDLE PATTERN

POURING HOLE

LOCATING DOWELS

WOOD

BLADE   CAST RESIN

BOTTOM HALF OF MOULDING BOX

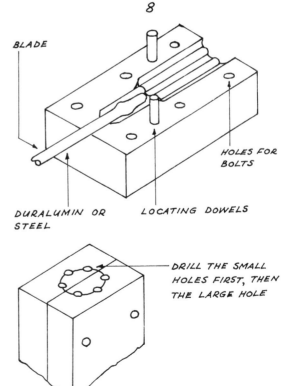

8

BLADE

HOLES FOR BOLTS

DURALUMIN OR STEEL   LOCATING DOWELS

DRILL THE SMALL HOLES FIRST, THEN THE LARGE HOLE

CONDIMENT SETS

MATERIALS – Copper, gilding metal or nickel silver 1 mm thick.

Develop the shape in one piece or in parts according to the design. Bend the shape in folding bars making sure that the joints seat well together.

Use soft iron binding wire to hold the joints firmly together. Well flux the joints and silver solder them. If more than one joint has to be soldered, use hard running silver for the first joint, easy running silver solder for the second and soft solder for the last.

The base is best set in so as to allow clearance for the filler plug. Make the plug from cork, flexible plastic or rubber or metal threaded to fit a corresponding hole in the base.

Experiment to find the best size to drill the holes for both the pepper and the salt, three, four or five holes for the pepper and one hole for the salt.

Round off all sharp corners and polish the surfaces.

Other possible materials

The materials from which the design is to be made will influence its shape and its construction.

Copper, brass or stainless steel tube, either left parallel or pressed together at the top end and the joint silver soldered.

Combinations of metal tube and wooden blocks.

Duralumin rod machined to shape.

Duralumin combined with cast resin or plastic.

Cast resin with a tubular insert.

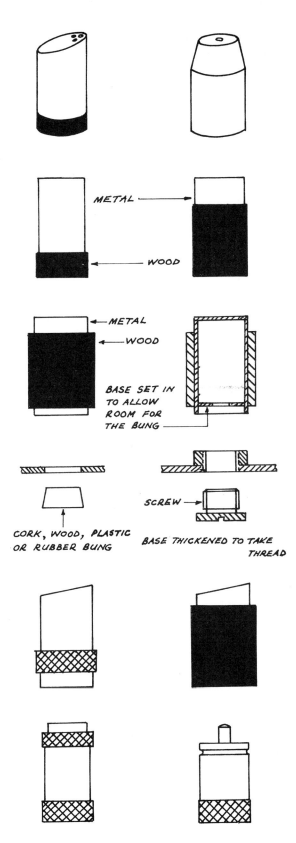

METAL

WOOD

METAL

WOOD

BASE SET IN
TO ALLOW
ROOM FOR
THE BUNG

CORK, WOOD, PLASTIC
OR RUBBER BUNG

SCREW

BASE THICKENED TO TAKE
THREAD

48

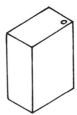
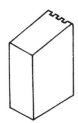

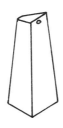

CONDIMENT SETS IN OTHER MATERIALS

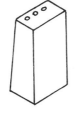

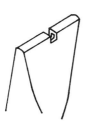

More examples on page 51

49

CONDIMENT SET

MATERIALS – Copper tube or aluminium alloy rod 25 or 30 mm diameter.
A dark coloured hardwood.
Plastic rod.

Make the main body from either tube or rod.

If tube is being used, silver solder on a top and a base, setting in the base to allow clearance for the head of the plug.

The base can be thickened by soldering a boss to it, thus providing a longer length of thread to take the plug.

If rod is being used, machine the base out of the solid and use an epoxy adhesive to fix on the top. Drill the pouring hole or holes.

Make the filler plug from plastic rod, threading it to fit the thread used in the base.

Drill and bore the hardwood to fit the outside diameter of the tube and machine the outside of it so that a thin wall is formed. Clean the surfaces and polish them.

Fit the hardwood sleeve to the body with an epoxy adhesive.

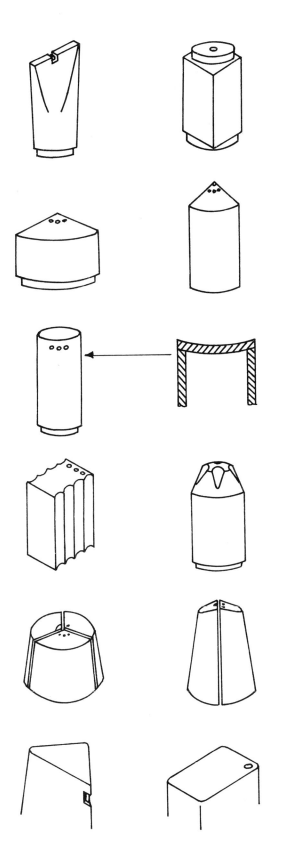

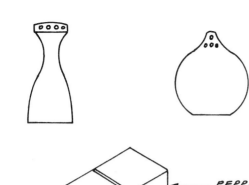

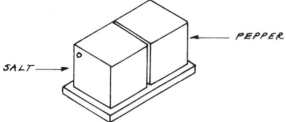

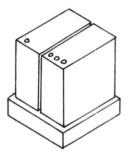

SALT →

→ PEPPER

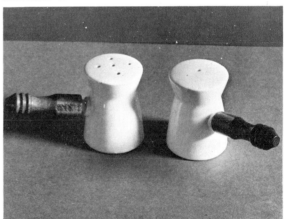

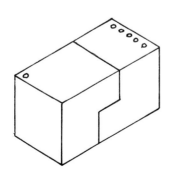

FISHING WEIGHTS

Design and make a steel mould for casting lead fishing weights of, say, 60 grams.

Design 1

a Leave the bore straight from the drill.
b Finish the base flat with a flat bottom drill, *or*
c Bore the base flat on a lathe.

A slight taper will aid extraction. Do this with a taper reamer or bore the taper on a lathe.

Obtaining the sizes by calculation

11.4 g is the weight of 1 cm³ of lead

60 g is the weight of $\dfrac{60}{11.4}$ cm³ of lead

$$= \frac{60\,000}{11.4}\text{mm}^3$$

$$\text{Vol.} = 5200 \text{ mm}^3 \text{ (approx.)}$$

Volume of Cylinder $= \pi r^2 \times h$, where h is the height and r the radius of the cylinder.
Let $r = 6$ mm

$$h = \frac{V}{\pi r^2}$$

$$h = \frac{5200 \times 7}{22 \times 36}$$

$$= 46 \text{ mm (approx.)}$$

Fishing weight and mould
(Example 5 opposite)

Design 2

Calculating the sizes

1 Make a drawing to scale on squared paper. Count the squares to find the cross-sectional area. This area multiplied by $\frac{1}{4}$ of the mean circumference is equal to the volume:

$$V = A \times \frac{C}{4}$$

or

2 Treat as a cylinder with a centre hole. Assume that the sides of the cylinder are straight. Omit the volume of the semi-circular projections to compensate.

or

3 Use a displacement jar.

$$\text{Volume of weight} = \pi h (R^2 - r^2)$$

$$h = \frac{V}{\pi (R^2 - r^2)}$$

Let $R = 18$ mm and $r = 6$ mm

$$h = \frac{5200}{\pi (18^2 - 6^2)}$$

$$h = \frac{5200 \times 7}{22(324 - 36)}$$

$$h = \frac{5200 \times 7}{22 \times 288}$$

$$h = 6 \text{ mm (approx.)}$$

Bore out the two mould halves. Form the projections by drilling into the metal a short way. Provision should be made for the mould to stand vertically whilst the pouring is being done.
Trap a piece of wire at the top of the cast before pouring to provide a fixing ring.

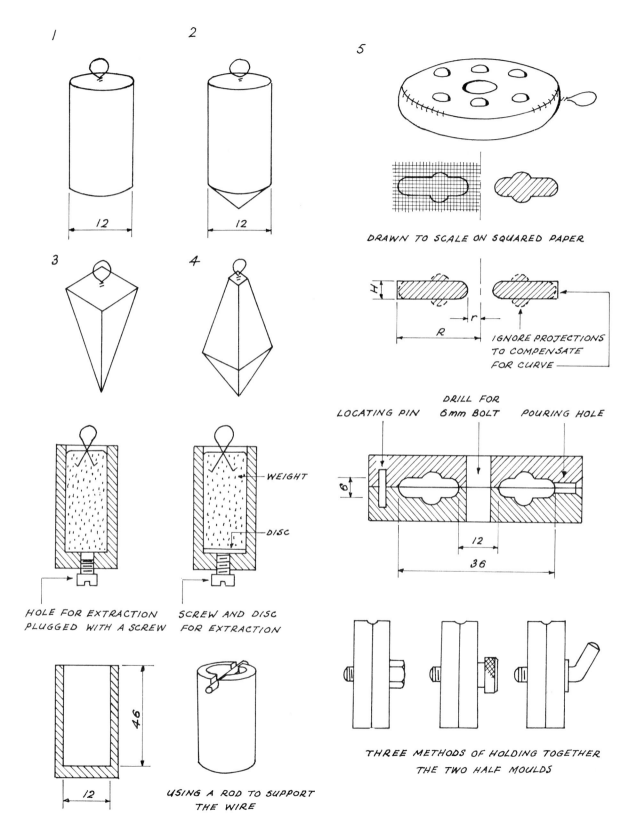

1

2

3

4

12

12

5

DRAWN TO SCALE ON SQUARED PAPER

H

r

R

IGNORE PROJECTIONS
TO COMPENSATE
FOR CURVE

WEIGHT

DISC

DRILL FOR
LOCATING PIN   6mm BOLT   POURING HOLE

6

12

36

HOLE FOR EXTRACTION
PLUGGED WITH A SCREW

SCREW AND DISC
FOR EXTRACTION

46

12

USING A ROD TO SUPPORT
THE WIRE

THREE METHODS OF HOLDING TOGETHER
THE TWO HALF MOULDS

53

TEA TRAY OR WINE TRAY

Design 1

BASE – 9 mm plywood, faced to form a suitable surface.

FRAME – 25 mm×1.5 mm aluminium alloy angle.

The frame is made in two halves with butt joints in the middle of the ends. The bends are made by cutting a 90° vee in the bottom face of the angle. Make the bend by hand.

When bent the corners will have a slight radius and no sharp corner.

The handles, fixed with screws or rivets, will reinforce the two joints in the frame.

Fix the base to the frame with contact adhesive and small screws.

Design 2

BASE – 9 mm plywood.

FRAME – 25 mm×1.5 mm aluminium alloy angle.

Make the frame from four lengths of alloy angle with a butt mitre joint at each corner.

Fix the base to the frame as above.

Design 3

BASE – 9 mm plywood.

FRAME ENDS – 1.5 mm aluminium alloy sheet.

FRAME SIDES – 6 or 8 mm diameter alloy rod.

Turn a spigot on the ends of the rods. Drill holes in the frame ends to suit the spigots.

Rivet over the ends of the spigots.

Fix the base to the frame with contact adhesive and screws.

Design 4

MOULD – Any suitable softwood.

TRAY – Polyester resin, pigment, catalyst and glass fibre mat.

Here a wide range of shapes and designs are possible. The handles can be moulded in one piece with the tray base.

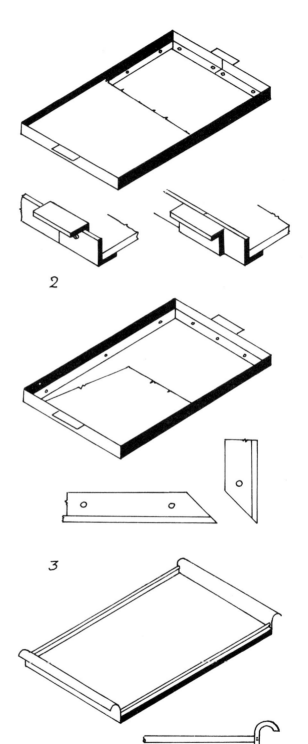

2

3

Handles

1 Handles formed in one with the ends of the tray. The handle shape is open to variation.

2 Handles applied separately. They can be made of aluminium alloy left bright or finished with an eggshell polyurethane paint. Or they can be made of Perspex, bent to shape by heating to about 150 °C and bending over a suitable former.

Fix the handles by means of 2 BA countersunk screws or 4 mm rivets. If the screws or rivets are positioned low down on the frame, they will be hidden by the tray base when it is set in.

3 As above but use aluminium alloy angle.

4 Form the handles by bending out the vertical faces of the alloy angle. Use 16 mm diameter alloy or plastic rod for the finger-grips. Face the rods in the lathe and drill and tap holes in the ends with a 2 BA thread. Assemble with 2 BA countersunk chromium-plated screws.

Facings for the base

1 Solid wood with a polyurethane finish.

2 Laminate sheeting, e.g. Formica, on 10 mm thick plywood.

3 Contact or Fablon sheeting on plywood. This is cheaper but not as durable as laminate.

4 Plastic tiles fixed to a plywood base with a tile cement. Obtain the tiles first and make the tray a size to fit the tiles.

5 Mosaic tiles. These are small coloured tiles with which a wide range of patterns can be built up.

6 Processed cork tiles. These are durable and 'kind' to the crockery. Fix with a contact adhesive.

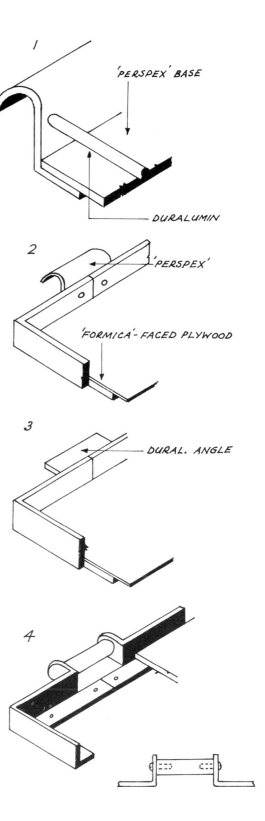

5 First principles in designing table and wall lamps

LAMPS

The modern electric lamp is a development of the old paraffin lamp and, too often, imitates its design and shape. The paraffin lamp demanded a base to hold the oil, a fitting to hold the glass and a stem to link the two together. An electric lamp does not have to satisfy the same demands and there is much scope for designs that break completely with rigid tradition.

Consider a table lamp from first principles – a light source (bulb) has to be supported and held in position above a plane surface in such a way that the light can be dispersed or directed to one spot efficiently. The problem is best solved by designing a complete unit and not an assembly of separate parts.

Other incidental problems are involved, such as provision for flex and switch etc.

The same principles will apply when designing a wall lamp.

Attaching the bulb holder

The base of the bulb holder is provided with an internal thread ½-inch diameter×26 t.p.i. (British Standard Brass Thread).

Two fittings are shown, which can be bought or made. Both are threaded to match the internal thread of the bulb holder.

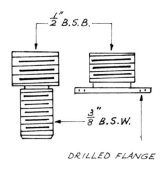

$\frac{1}{2}''$ B.S.B.

$\frac{3}{8}''$ B.S.W.

DRILLED FLANGE

BRASS FITTING

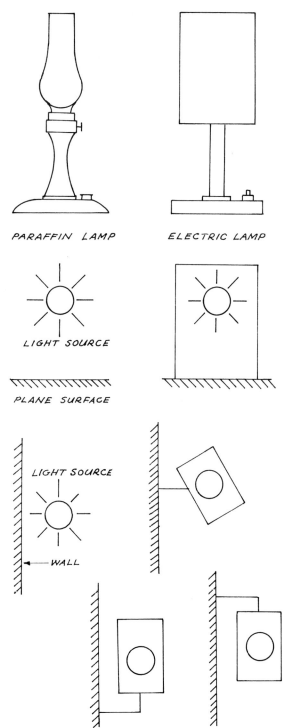

PARAFFIN LAMP

ELECTRIC LAMP

LIGHT SOURCE

PLANE SURFACE

LIGHT SOURCE

WALL

Bulb holders

Bulb holders can be obtained with or without a switch. Both are threaded internally with a ½-inch diameter×26 t.p.i. thread (BSB). The top of the holders are threaded with a 1⅛-inch diameter×26 t.p.i. thread and two screwed collars for the attachment of a shade.

If a plain bulb holder, without a switch, is used, a press-button switch can be wired into the circuit.

Fitting the switch

Connect the switch by breaking the live lead. Use 3-core flex and fix the:

Green/yellow lead to 'E', the earth terminal of the plug.
Brown lead to 'L', the live terminal of the plug.
Blue lead to 'N', the neutral terminal of the plug.

Fix the live and neutral leads of the flex to the lamp holder and the earth lead to any convenient metal part of the lamp.

The shade

Make a shade frame from 2 or 3 mm diameter steel rod, e.g. welding rod, and braze the joints. Incorporate a ring of 1⅛-inch (28.6 mm) inside diameter in the frame to suit the thread on the bulb holder. If acrylic sheet, e.g. Perspex, is used for the lamp or the shade, build the ring unit in at any suitable position.

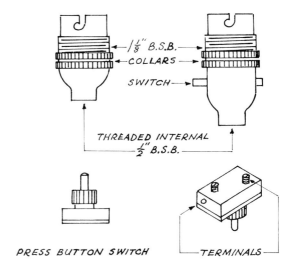

$1\frac{1}{8}$" B.S.B.
COLLARS
SWITCH

THREADED INTERNAL
$\frac{1}{2}$" B.S.B.

PRESS BUTTON SWITCH   TERMINALS

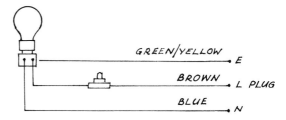

GREEN/YELLOW   E
BROWN   L PLUG
BLUE   N

THE ELECTRIC WIRING

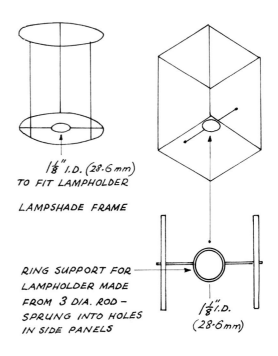

$1\frac{1}{8}$" I.D. (28·6 mm)
TO FIT LAMPHOLDER

LAMPSHADE FRAME

E

RING SUPPORT FOR LAMPHOLDER MADE FROM 3 DIA. ROD – SPRUNG INTO HOLES IN SIDE PANELS

$1\frac{1}{8}$" I.D. (28·6mm)

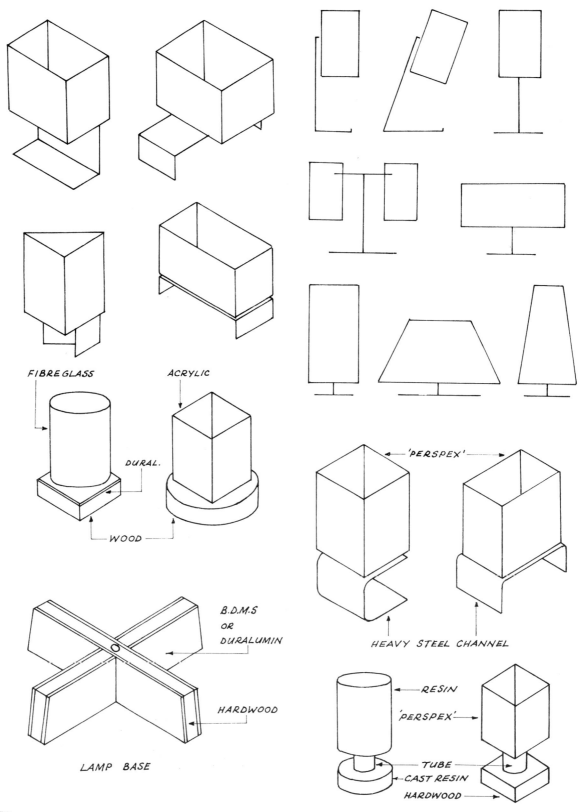

FIBREGLASS

ACRYLIC

DURAL.

WOOD

'PERSPEX'

HEAVY STEEL CHANNEL

B.D.M.S
OR
DURALUMIN

HARDWOOD

LAMP BASE

RESIN

'PERSPEX'

TUBE

CAST RESIN

HARDWOOD

MAKING A LAMP FROM ACRYLIC SHEET

1 Cut the rectangular pieces of acrylic sheet to size, making sure that the edges are square.

2 Bevel the mating edges by filing them to an angle of 45°. The use of a hardwood jig or fixture to support the material will help to maintain the straightness of the edges and their correct angle, as shown in the drawing.

3 Drill a series of holes in the sides so that they form an attractive pattern. Back these holes with coloured translucent acrylic sheet.

4 Assemble the four pieces using Tensol No. 6 cement to make the joints. Hold the joints together with self-adhesive tape until the adhesive has set.

5 Make the base from a good quality hardwood, rebating the top edges so that the acrylic assembly sits onto it.

6 Drill a hole for the brass fitting and screw in this fitting to hold the bulb holder. Drill a suitably sized hole for the flex.

7 Finish the base with wax or polyurethane paint.

A top may be fitted, made from clear or translucent acetate, drilled to form a pattern and to allow heat to escape.

Shades can also be made using fine gauze or expanded metal instead of acrylic sheet. The surfaces can be left plain or decorated with polyester resins.

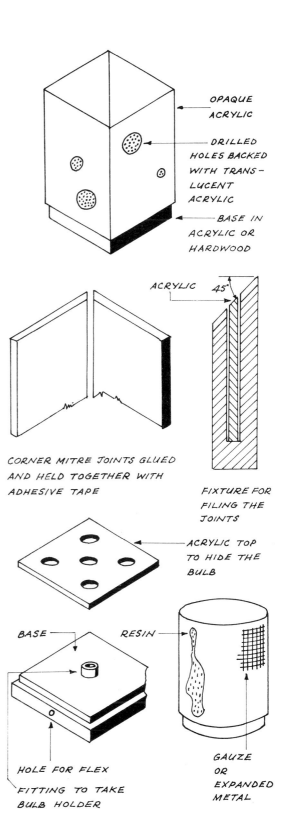

OPAQUE ACRYLIC

DRILLED HOLES BACKED WITH TRANS- LUCENT ACRYLIC

BASE IN ACRYLIC OR HARDWOOD

ACRYLIC 45°

CORNER MITRE JOINTS GLUED AND HELD TOGETHER WITH ADHESIVE TAPE

FIXTURE FOR FILING THE JOINTS

ACRYLIC TOP TO HIDE THE BULB

BASE

RESIN

HOLE FOR FLEX

FITTING TO TAKE BULB HOLDER

GAUZE OR EXPANDED METAL

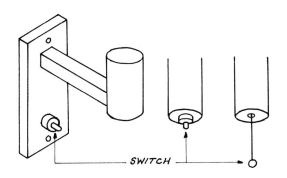

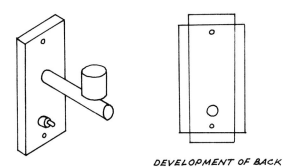

DEVELOPMENT OF BACK

WALL LAMPS

Make the backplate from either (a) sheet metal bent to shape with braze joints or (b) wood chiselled out at the back to accommodate the wiring and the switch.

The switch can be fitted to the backplate or to the lamp itself. Use a press-button switch or a cord switch.

Remember to break the *live* lead of the flex to wire in the switch.

Swivelling shades

1 The shade and the stem swivel on the backplate, a short coiled spring providing the necessary tension. A split pin and a washer act as a stop for the spring. The sleeve can be made of metal, wood or plastic.

2 This type of fitting is more suitable for desk lamps and workshop lamps. A lever-type handle locks the unit in position.

3 This locking device is similar to the one used on standard surface gauges. The design is more suitable for a shade made of metal or glass-reinforced resins.

SWIVELLING SHADES

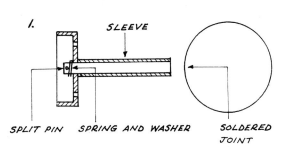

SPLIT PIN   SPRING AND WASHER   SOLDERED JOINT

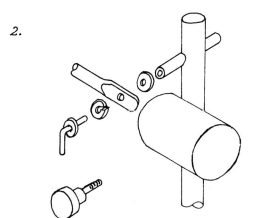

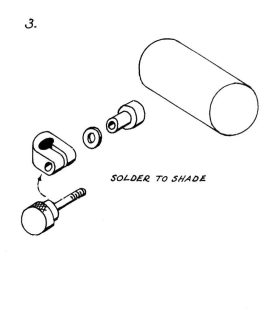

SOLDER TO SHADE

60

PROBLEM 1: FITTING A TUBE TO A WOODEN BASE

Solution 1

Make a plug to fit inside the tube. Drill a hole through the plug to take the flex.
Turn down one end of the plug and thread it to take a standard nut. Hard solder or soft solder the plug onto the tube.

Solution 2

Make a metal flange and bore it to fit the outside diameter of the tube. Drill 3 or 4 holes in the rim so that it can be screwed onto the wooden base. Hard solder or soft solder the flange to the tube.

Solution 3

Drill or bore a hole part way through the base of such a size that the tube will be a push fit in it. Fasten the tube to the base by means of an epoxy resin

Making the base heavy enough

If the lamp lacks stability because the base is too light in weight, machine a dovetail-shaped recess in the underside of the base. Fill this space with molten lead.

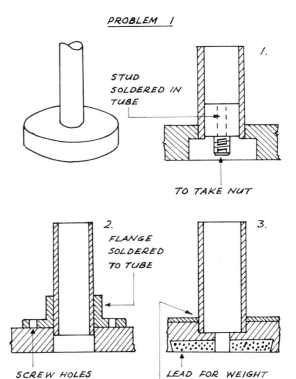

PROBLEM 1

1.
STUD SOLDERED IN TUBE

TO TAKE NUT

2.
FLANGE SOLDERED TO TUBE

3.

SCREW HOLES

LEAD FOR WEIGHT

EPOXY ADHESIVE

PROBLEM 2: FIXING A BRASS FITTING

Solution

Turn a plug, with or without a shoulder, to fit the inside diameter of the tube.
Drill and tap the plug $\frac{3}{8}$ inch BSW. Hard solder or soft solder the plug into the tube.

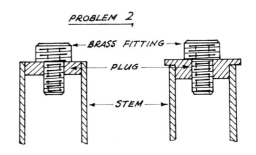

PROBLEM 2

BRASS FITTING

PLUG

STEM

BENDING AND FORMING ACRYLIC SHEET

Acrylic sheet is a thermoplastic, which means that it will soften when heated to such an extent that it can be bent under light pressure. The shape is retained when the plastic is cooled.

Bending

Bending is best performed by local heating rather than by heating a large area. A strip heater, bought or made, will heat the acrylic along a narrow length. The temperature needed for bending is about 150 °C.

A strip heater can be made from a length of Nichrome resistance wire, wired to the 12-volt terminal of a transformer. Find the exact length of wire by trial and error. Use a piece of wood for a base and cut a groove along its length. Mix some fireclay into a paste and fill the groove with it. Press a length of steel rod into the surface of the clay so that it forms a semi-circular recess into which the wire will seat.

Heat both sides of the sheet, bend to shape in a forming jig and allow to cool slowly.

Forming

Acrylic sheet, when heated, can be shaped most effectively on formers. Make a wooden former and line the forming face with soft cloth. Heat the plastic in a gas or an electric oven to a temperature of 150 ° C, supporting the sheet on a cloth-lined flat surface. Heating time – 10 minutes for every 3 mm of thickness. Use cotton gloves to hold the hot sheet, both to protect the hands and to avoid marking the sheet.

Lay the sheet over the former and pull it firmly into position by means of a length of strong cloth fitted with wooden strips. Maintain the pressure until the sheet has cooled slowly. Quick cooling encourages distortion.

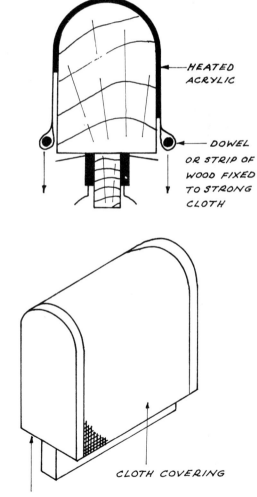

HEATED ACRYLIC

DOWEL OR STRIP OF WOOD FIXED TO STRONG CLOTH

CLOTH COVERING

WOODEN FORMER

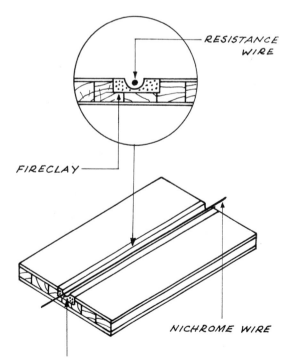

RESISTANCE WIRE

FIRECLAY

NICHROME WIRE

GROOVE IN BASEBOARD FILLED WITH FIRECLAY

EXAMPLES OF TABLE AND WALL LAMPS

6 Stool design

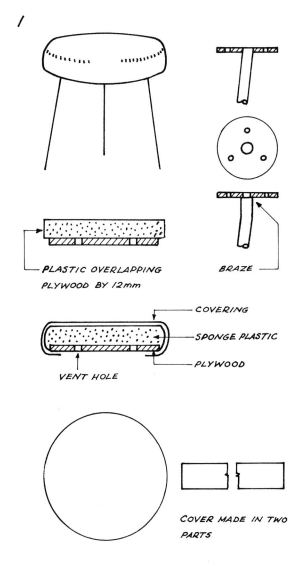

PLASTIC OVERLAPPING PLYWOOD BY 12mm

BRAZE

COVERING

SPONGE PLASTIC

PLYWOOD

VENT HOLE

COVER MADE IN TWO PARTS

STOOLS 1

LEGS – 9 mm diameter BDMS rod or 12 mm diameter BDMS tube.

Braze the legs to the fixing plates.

Problem. How to drill the holes at an angle in the fixing plates.
Solution. **(a)** Make a jig to drill the centre holes.
(b) Drill the holes square to the plates and bend the legs.

UPHOLSTERED TOP – Sponge plastic on plywood or blockboard, covered with fabric or plastic sheeting, e.g. vinyl.

Drill four or six holes, 12 mm diameter, in the baseboard for the release of air. Cut the sponge plastic larger than the baseboard so that it projects about 12 mm all round. Fix it to the baseboard with a little impact adhesive.
Make the cover in two pieces and stitch the pieces together.
Fix the cover material to the base with upholstery tacks or staples.

Alternative construction

Instead of fixing plates, make a triangular underframing from 12 mm×3 mm steel angle. Form the shape by cutting out vees at 120° angles where each bend is to be made. Form a butt joint in the middle of one of the sides. Braze the butt joint. Drill three holes in the frame to take the legs.
Braze the legs in position at the same time as brazing the corner joints.

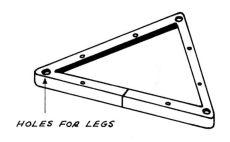

HOLES FOR LEGS

ALTERNATIVE FIXING USING STEEL ANGLE

120°

FRAME BEFORE BENDING

STOOLS 2a

LEGS – 9 mm diameter BDMS rod.

TOP BEARERS – BDMS strip 3 mm thick.

TOP – Sponge plastic on a plywood base, all covered with fabric or vinyl sheeting.

Fix the bearers either on top of or underneath the end frames. Use a jig for bending the legs to make sure that both end frames are the same.

Rivet the bearers to the end frames and braze these joints for extra strength.

Fix the framing to the seat by screwing either through the bearers or through the frames.

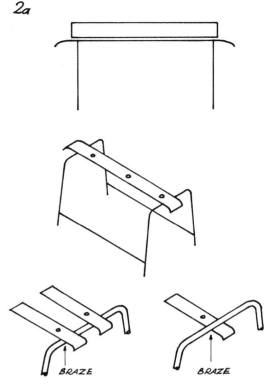

2a

BRAZE   BRAZE

BEARERS ON TOP OF LEGS   BEARER UNDER LEGS

STOOLS 2b

LEGS – 12 mm square steel tube.

RAILS – 8 or 10 mm diameter BDMS rod.

TOP – as above.

Cut vee notches in the tubes where the bends are to be. Bend at these points and braze the joints.

Braze the rails into the legs.

Fix the top by screwing through the end frames.

STOOLS 2c

LEGS – Steel tube 12 mm square or 12 mm diameter.

RAILS – 8 or 10 mm diameter BDMS rod.

BEARERS – BDMS strip 3 mm thick.

TOP – as above

Fix the legs to the bearers by any of the methods shown and braze the joints.

The rails fit into holes in the legs and the joints are brazed.

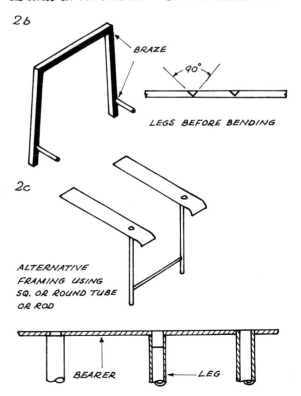

2b

BRAZE

90°

LEGS BEFORE BENDING

2c

ALTERNATIVE FRAMING USING SQ. OR ROUND TUBE OR ROD

BEARER   LEG

STOOLS 3

LEGS – 10 mm diameter BDMS rod or 12 mm
    diameter steel tube.
BEARERS – 8 mm diameter BDMS rod.
SEAT – Seagrass.

Bend the legs in a suitable jig to ensure uniformity
of the bends.
Drill holes in the legs to accommodate the bearers.
Keep the bearers their full diameter or turn spigots
on the ends of the bearers.
Braze all the joints.
The top is woven in seagrass – for details of sea-
grass weaving see page 99.

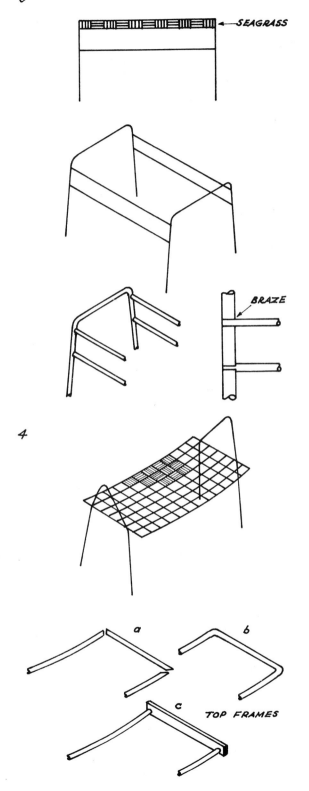

STOOLS 4

4

LEGS – BDMS rod 10 mm diameter.
TOP FRAME – BDMS rod 10 mm diameter.

Bend the legs in a suitable jig.
Choose one of three methods of forming the top
frame.

(a) Use four separate pieces of metal, mitred at
the corners and then brazed.
(b) Use one length of metal. Bend the corners and
make the butt joint in one of the sides. Braze the
joint.
(c) Use BDMS strip for the ends and BDMS rod
for the sides.
Drill the strips to take the sides and braze the
joints.

TOP – Seagrass – for details of seagrass weaving see
    page 99.

STOOLS 5

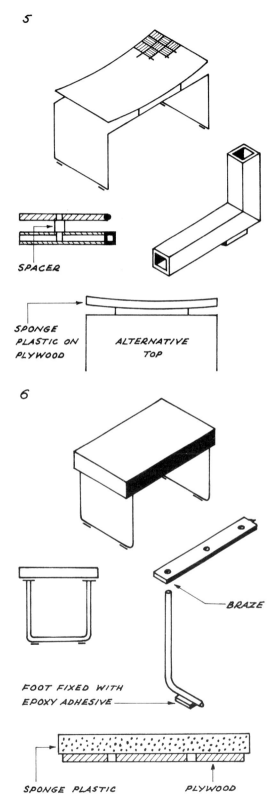

MAIN FRAME – 12 mm square steel tube.

TOP FRAME – 8 mm diameter BDMS or 8 mm BDMS square.

SPACERS – 8 mm diameter BDMS rod.

TOP – Woven seagrass or covered sponge plastic.

Make the main frame from either:
(a) short lengths of tube, mitred at the corners and brazed, or
(b) one long length of tube, cut with 90° vees at the bends and brazed at the joints.

Make up the top frame with mitred joints at the corners, brazed.

Turn spigots on both ends of the spacers. Fit them between the main frame and top frame. Rivet the joints and then braze. For details of the upholstered top, see Stools 8.

SPACER

SPONGE PLASTIC ON PLYWOOD

ALTERNATIVE TOP

STOOLS 6

FRAMES – 10 or 12 mm diameter BDMS rod.
BEARERS – 16 mm×4 mm BDMS strip.

Fix the end frames to the bearers by either:
(a) drilling the bearers with holes the same diameter as the frames and brazing the joints, or
(b) drilling 3 mm holes in the bearers and in the top end of the frames and locating the two parts with 3 mm rivets. Braze the joints.

TOP – Solid hardwood suitably finished, blockboard faced with quilted vinyl, or sponge plastic on plywood covered with a serviceable material.

Fix metal, plastic or rubber feet to the underside of the frames using an appropriate adhesive.

BRAZE

FOOT FIXED WITH EPOXY ADHESIVE

SPONGE PLASTIC   PLYWOOD

67

STOOLS 7

FRAME – 12 mm diameter BDMS rod or 16 mm
    diameter×18 g steel tube.
TOP – Sponge plastic on plywood.
BEARERS – 12 mm diameter BDMS rod.

Bend the end frames in a jig or a bending machine.
File a curve on the ends of the bearers to fit the
curve of the end frames and drill a hole in both
bearer and frame to take a locating pin.
Braze the joints.
Fix the top by screwing up through the frame.

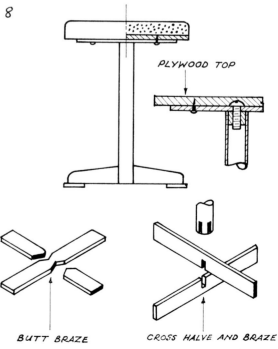

7

SCREW OR
RIVET

STOOLS 8

8

TOP – Sponge plastic, 25 or 50 mm thick, on 12 mm
    plywood. Covered with fabric or vinyl cloth.
STEM – Steel tube 30 or 35 mm diameter.
BEARERS – 25 mm×4 mm BDMS strip.
FEET – 25mm×6 mm BDMS strip.

For the details of the top, see Stools 1.
To fix the stem to the top, machine a plug to fit the
inside diameter of the tube and drill and tap it M8.
Braze the plug into the tube and screw the bearers
to it. Screw through the bearers into the plywood
base of the upholstered top.
Join the feet by means of a cross halving joint and
slot them into the stem. Braze the joint.

PLYWOOD TOP

BUTT BRAZE   CROSS HALVE AND BRAZE

STOOLS 9

LEGS – BDMS rod 12 mm diameter.

RAILS and BEARERS – BDMS rod 8 or 10 mm
diameter.

TOP – Sponge plastic on 12 mm thick plywood or
chipboard, or shaped in G.R.P.

Turn spigots on the ends of the bearers and drill
holes in the leg frames the same diameter as the
spigots. Assemble by riveting or brazing or both.
Shape the rail ends to fit the curvature of the legs
and drill them lengthwise (longitudinally) to take a
5 or 6 mm diameter dowel. Drill the legs in the
correct position to accommodate the dowels and
braze all the rail joints.

Sponge plastic top. Construct in the same way as
Stool 1.

GRP top. Make a suitably shaped former or
pattern from 3 or 4 mm thick plywood bent over
battens on a wooden baseboard. Use 4 or 5 layers
of $1\frac{1}{2}$ oz. mat with the resin pigmented to the
required colour. Machine the heads of 4
hexagonal-headed bolts or roundhead screws to
make the heads thin and mould them into the resin
top in such a way that they line up with the bearers.
Drill the bearers 6 mm clearance and fix the top to
the underframing by means of 6 mm nuts.
Finish off the ends of the legs with rubber feet or
plastic end cups.

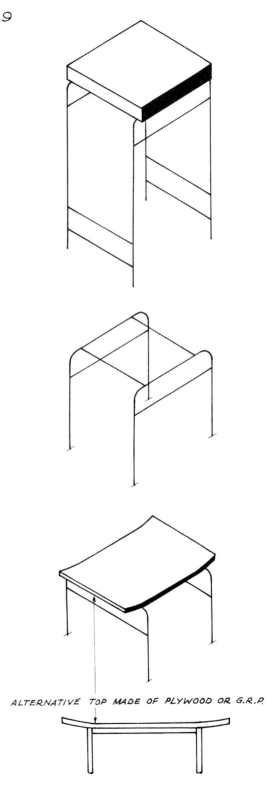

ALTERNATIVE TOP MADE OF PLYWOOD OR G.R.P.

STOOLS 10

LEGS – Semi-bright steel tube 16 mm diameter ×
   18 g, or BDMS rod 12 mm diameter.
RAILS – BDMS rod 8 mm diameter.
BEARERS – BDMS strip 25mm × 5 mm.

The constructions shown in the drawings obviate
the need for drilling holes at an angle in the
bearers.

Using a long length of metal, bend the bearers cold
and then cut them to length. Drill holes in the
bearers to take either the 16 mm diameter tube or
the 12 mm diameter rod. Braze these joints when
the assembly is complete.

Use either of the methods shown to fix the rails to
the legs and braze the joints.

Form the top from three layers of 4 mm thick
plywood, by gluing them together and cramping
them to the bearers or by cramping them over
wooden blocks, as shown.

Drill about six 12 mm diameter holes in the shaped
plywood to allow air to escape when the stool is sat
on.

Complete the upholstery as detailed for Stool 1.

Finish the ends of the legs with rubber feet or
plastic end cups.

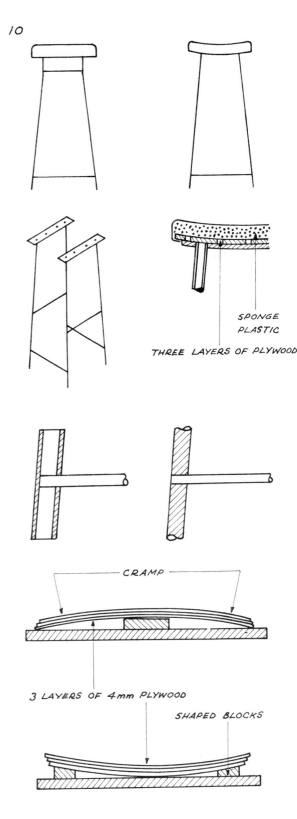

SPONGE PLASTIC

THREE LAYERS OF PLYWOOD

CRAMP

3 LAYERS OF 4mm PLYWOOD

SHAPED BLOCKS

70

STOOLS 11

MAIN FRAME – BDMS rod 10 mm diameter.
RAILS – BDMS rod 8 mm diameter.
TOP – Seagrass or upholstery cord.

To shape the main frame, heat up the metal and
bend in the correct places whilst the metal is hot.
Braze the two legs to the upper part of the frame
using 3 mm diameter rivets to locate the parts.
In the same way, locate the rails and braze the
joints.
For details of seagrass weaving, see page 99.

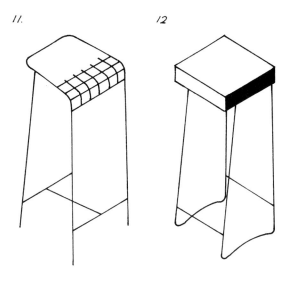

STOOLS 12

FRAME – BDMS rod 10 mm diameter.
RAILS – BDMS rod 8 mm diameter.
BEARERS – BDMS strip 20 mm×4 mm.
TOP – Sponge plastic on plywood or hardwood.

Make the frame by first forming the base curves
and then making the bends by using a jig.
Locate the rails on the legs with 3 mm diameter
rivets and braze the joints. Position holes in the
bearers to line up with the frame and drill them 10
mm diameter. Braze the joints.
For details of the top, see page 99.
Fix the top to the frame by screwing through the
bearers.

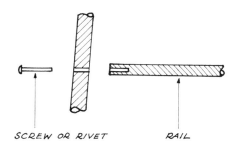

SCREW OR RIVET   RAIL

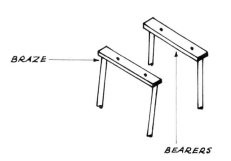

BRAZE

BEARERS

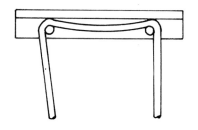

BENDING JIG

STOOL DESIGN

7 Table design

TABLES 1

/.

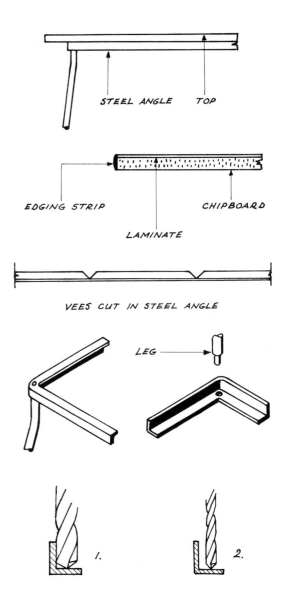

STEEL ANGLE   TOP

EDGING STRIP   CHIPBOARD

LAMINATE

VEES CUT IN STEEL ANGLE

LEG →

/.   2.

FRAME – Steel angle 16mm×16mm×3 mm.
LEGS – BDMS rod 10 mm diameter.
TOP – Chipboard or block board 16 mm thick faced
    with a laminate, e.g. Formica.

Cut the steel angle to size and mark out for the
bends. Cut out 90° vees where the bends are to be
and form a butt joint in the middle of one of the
sides. Braze the butt joint and the corner joints.
To fix the legs, using a 10 mm drill, drill into the
corner of the steel angle and once the drill has
started to cut, withdraw the drill and then drill
through with a 6 mm drill. Turn a spigot 6 mm
diameter on the ends of the legs, rivet to the frame
and silver solder. Alternatively, the brazing of the
steel angle frame and the legs can be done in one
operation. The legs can be bent before or after
assembly using a length of tubing to provide leverage.
Make the top from 16mm thick chipboard or
blockboard. Cover the top face with Formica using
an impact adhesive.
Trim the edges flush and face the edges with flexible plastic strip or aluminium alloy strip or wood.

TABLES 2

LEGS – BDMS rod 12 mm diameter.
TOP FRAME – Steel angle 16mm×16mm×3 mm.
TOP – Plate glass or tiles on 12 mm plywood.

Make the top frame and the legs as shown in Chair 1, but position the steel angle in reverse so as to form a seating for the top.
Tiles are not always exact for size, so obtain or make the tiles first before deciding on the exact size of the top. Bed in the tiles with a tile cement.
Use plastic or rubber feet for the ends of the legs.
Finish all metal surfaces with a satin finish paint.

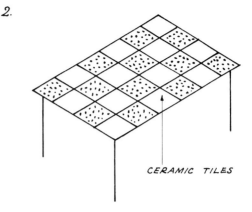

2.

CERAMIC TILES

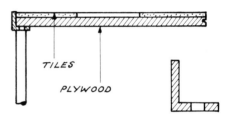

TILES

PLYWOOD

TABLES 3

3.

FRAME – Steel tube 12 to 25 mm square, according to the size of the table.
TOP – Plywood 12 or 16 mm thick faced with Formica.

Form the end frames from two lengths of square tube, notched out at 90° for the bends. Bend the two corners of each frame at 90°, fix together with clamping plates and braze the joints.
Locate the rails by either of the methods shown, clamp together and braze the joints. Finish the frame with a satin finish paint. Make or obtain plastic or rubber feet for the ends of the legs.
Face the top with Formica and edge the top with either Formica strip or flexible plastic strip or wood or aluminium alloy. Screw through the frame to fix the top.

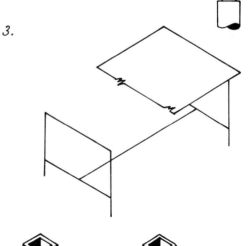

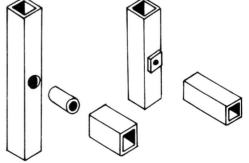

TABLES 4

LEGS – BDMS rod 12 mm diameter.
LEG BRACES – BDMS rod 6 mm diameter.
TOP FRAME – Steel angle 16 mm×16 mm×3 mm.
TOP – Plywood or chipboard 12 or 16 mm thick, with a suitable facing.

Construct the frame and the legs in the same way as Table 1.

Drill the frame at the corners with a 12 mm drill to position the hole and then drill through with an 8 mm drill. Turn an 8 mm spigot on the end of each leg. Rivet the legs to the frame and braze each joint.

Make the leg braces either straight or curved. Make a template for the braces if you decide to have them curved. Rivet the braces to both legs and frame and braze the joints.

Construct the top as shown in Table 1.

4.

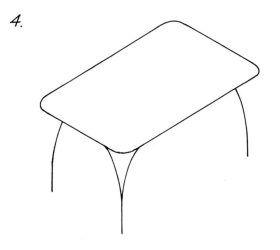

PIN AND BRAZE

TABLES 5

MAIN FRAME – Steel tube 12 mm square.
SHELF – Steel tube 12 mm square and 6 mm rod or 8 mm and 6 mm diameter BDMS rod.

Make the side frames from one length of tube, notching out at 90° for the corner bends. Use clamping plates to hold the joints together and braze them.

Drill the cross rails of the frame either the same size as the rod of the shelf or to take a tubular dowel if square tube is being used. Braze the joints. Drill holes to take the shelf cross pieces and braze the joints.

Use any of the previous methods for constructing the top.

Fit plastic or rubber feet to the ends of the legs.

5.

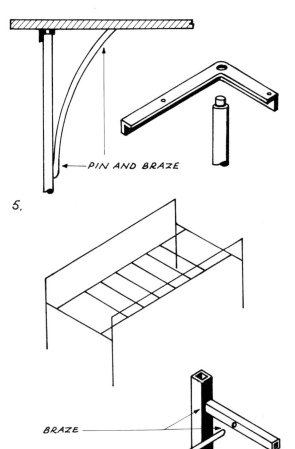

BRAZE

TABLES 6

SIDE FRAMES – Steel tube 38 mm×12 mm rect.

TOP RAILS – Steel tube 12 mm diameter.

BOTTOM RAILS – Steel tube 12 mm square.

TOP – Plate glass 6 mm thick or acrylic sheet 9 or 10 mm thick.

BOTTOM SHELF – Plywood 12 mm thick.

Make each side frame from one length of rectangular tube. Notch out at 90° for the bends and bend at these points.

Hold the joints together with clamping plates and braze the joints.

Drill holes in the side frames for the top rails and braze the joints.

Locate the bottom rails with tubular dowels inserted into the square tube and drill holes in the frames the same size as the dowels. Braze the joints.

Support the glass or acrylic sheet by means of rubber or plastic studs inserted into holes drilled in the side frames.

Face and edge the plywood shelf with veneer or suitable laminate. Fix the shelf by screwing up through the bottom rails.

Fit castors to the four legs.

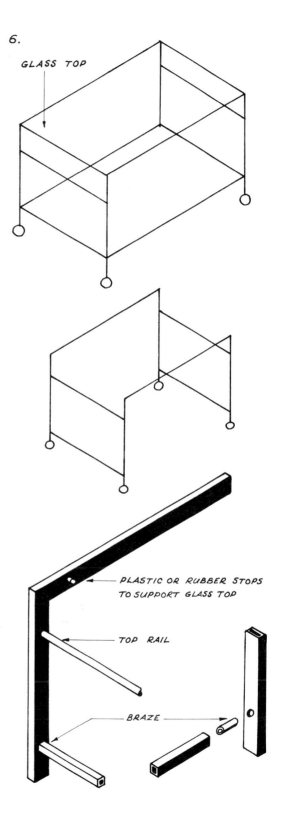

6.

GLASS TOP

PLASTIC OR RUBBER STOPS
TO SUPPORT GLASS TOP

TOP RAIL

BRAZE

TABLES 7

UNDERFRAMING – Steel tube 18 mm square.

TOP – Blockboard 16 mm thick, faced with a laminate.

Bend each end frame from one length of tube, by notching out a 90° vee where the bends are to be made.

Make a butt joint in the middle of the base of the frame and braze all the joints.

Locate the centre bearer by means of a square plug or a tubular dowel and braze.

Use clamping plates to hold the joints in place whilst brazing.

Fix rubber or plastic feet under the end frames by using an epoxy adhesive or screws or both.

7.

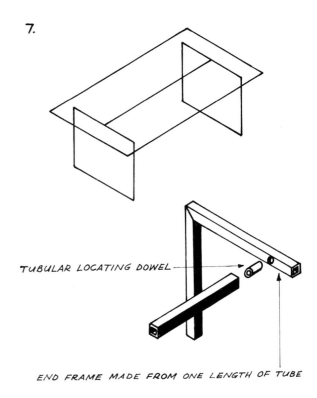

TUBULAR LOCATING DOWEL

END FRAME MADE FROM ONE LENGTH OF TUBE

TABLES 8

FRAME – Steel tube 18 mm square.

TOP – Blockboard 16 mm thick, faced and edged with a suitable material.

Construct the frame by using tubular dowels at each joint. Hold the joints together by means of clamping plates during the brazing.

Plug the exposed tube ends with wood or metal, fixing the plugs with an epoxy adhesive or plug the ends with a resin paste.

Fix four feet under the legs to give stability.

Screw through the underframing to fit the top.

8

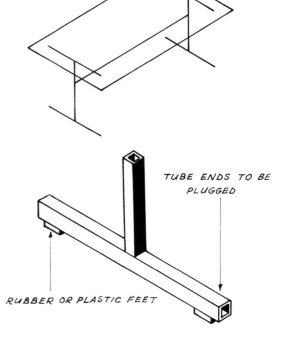

TUBE ENDS TO BE PLUGGED

RUBBER OR PLASTIC FEET

TABLES 9

UNDERFRAMING – BDMS rod 10 or 12 mm diameter according to the size of the table.

TOP – Blockboard faced with a laminate and edged with plastic strip.

Make the bend in each one of the legs using a jig. Make the brazed joints either by overlapping the joints, wiring them in position and then trimming the waste off after brazing, or by butting the joints and holding them in position with screws or rivets. Fix the top by either brazing a circular fixing plate to the top of the legs or by screwing down through the top into the tops of the legs. If the latter method is used, fix the top to the frame before facing it with laminate, so that the screw heads will be hidden by the facing.

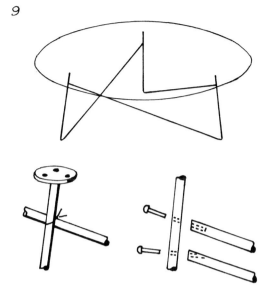

9

WIRE TOGETHER, BRAZE AND TRIM OFF WASTE   PIN AND BRAZE

TABLES 10

UNDERFRAMING – BDMS rod 10 or 12 mm diameter.

TOP – Blockboard 16 mm thick, faced and edged as desired.

Construct the underframing by one of the above methods.
Finish the ends of the legs with rubber or plastic feet.

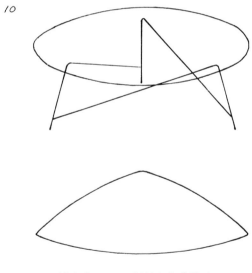

10

ALTERNATIVE SHAPE FOR TOP

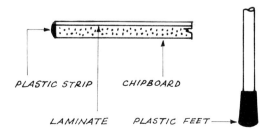

PLASTIC STRIP   CHIPBOARD

LAMINATE   PLASTIC FEET ⟶

TABLES 11

UNDERFRAMING – BDMS rod 12 mm diameter.
TOP – Chipboard or plywood 12 mm thick, faced
    with a laminate.

This design is suitable for a garden table.
Use a bending jig to bend the legs to ensure that
the bends are the same on each leg.
Fix the top by screwing through the underframing.
Fit plastic ball feet or plastic cups to the ends of the
legs.

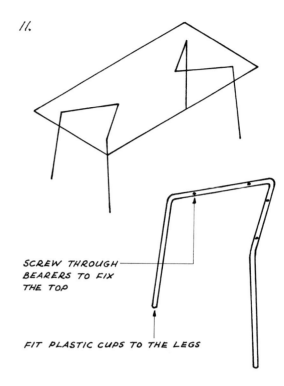

11.

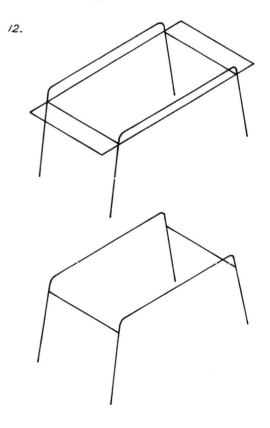

SCREW THROUGH
BEARERS TO FIX
THE TOP

FIT PLASTIC CUPS TO THE LEGS

TABLES 12

FRAME – BDMS rod 10 or 12 mm diameter or steel
    tube 12 mm diameter.
TOP – Chipboard or plywood 12 mm thick or block-
    board 15 mm thick.

Make the bends by using a jig to ensure uniformity.
Locate the bearers by means of small screws or
rivets and braze the joints.
Fix the top by screwing through the bearers or
braze fixing plates to the bearers and screw
through them into thc top.
If chipboard is used for the top, glue wooden
dowels into the chipboard where the screws are
positioned in order to give the screws a better hold.
The chipboard itself will not provide a firm fixing.

12.

TABLES 13

FRAMING – BDMS rod 12 mm diameter or steel tube 12 mm diameter.
TOP – Blockboard 12 mm thick or glass 6 mm thick.

Use a jig to make the bends.
Position the bearers to the end frames with screws or rivets and braze the joints.
Fix the top by screwing through the bearers, gluing wooden dowels in the chipboard where the screws will be, in order to give a stronger fixing.
Face the chipboard with a laminate or ceramic tiles.
Fix a glass top by resting it on rubber inserts glued into holes drilled in the bearers.
Finish off the leg ends with suitable feet.

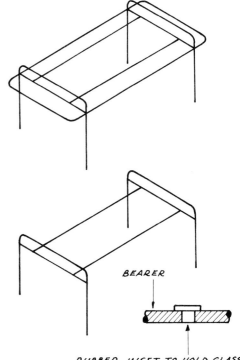

13.

BEARER

RUBBER INSET TO HOLD GLASS

TABLES 14

FRAMING – Steel tube 16 mm square.
TOP – Blockboard 16 mm thick, suitably faced.

Make the frame from two lengths of tubing and construct the joints as in the previous examples.
Locate the two parts of the frame with tubular dowels and braze the joint.
Fix the top by either plugging the top ends of the frame and drilling and tapping the plugs to take screws or by brazing plates to the top ends of the frame and screwing through the plates into the top.
If the former method is used, face the top after it has been fixed in position. Plug the bottom of the leg with a resin paste or fit plastic feet.

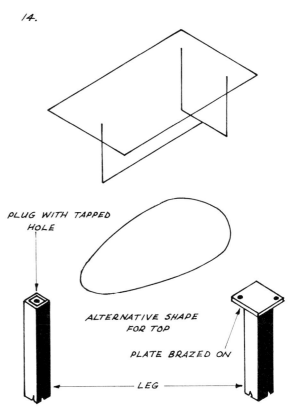

14.

PLUG WITH TAPPED HOLE

ALTERNATIVE SHAPE FOR TOP

PLATE BRAZED ON

LEG

TABLES 15

FRAME – BDMS or duralumin rectangular strip and angle.
TOP – Blockboard with a hardwood surround.

(a) Duralumin frame

Fix the strip to the inside or the outside of the angle and screw together either with countersunk screws from the inside or with round-headed chromium-plated screws from the outside. An epoxy resin applied to the joint will give extra strength.

(b) Steel frame

Construct as above but braze the joints.
Braze or screw small angle brackets to the sides of the frame and screw through them to fix the top. Make the top from blockboard with a mitred hardwood surround. Cover the blockboard with vinyl sheeting or leather.

TABLES 16

FRAME – BDMS rod 12 mm diameter.
TOP – Glass or acrylic sheet.

Construct the frame by brazing the rails to the legs. A problem will have to be solved here – the method of holding the parts together for brazing. Fix the top by any one of the following methods:

(a) Screw and/or braze small angle brackets to the tops of the legs and drop the top in.
(b) Fit a metal or plastic dowel into the top of each leg and drill the top to fit. Insert a rubber or soft plastic washer between the top and the legs.
(c) Fit plastic suckers into the tops of the legs.

15.

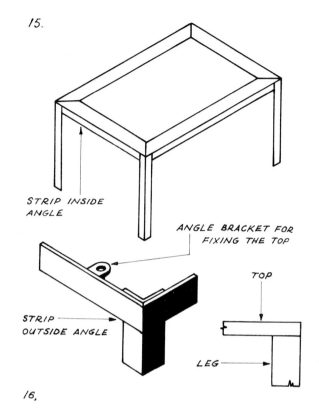

STRIP INSIDE ANGLE

STRIP OUTSIDE ANGLE

ANGLE BRACKET FOR FIXING THE TOP

TOP

LEG

16.

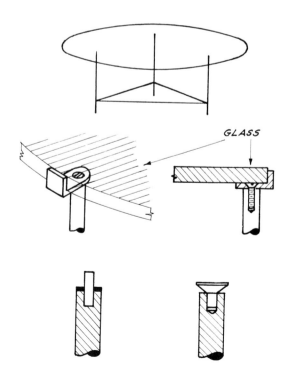

GLASS

82

EXAMPLES OF TABLE DESIGN

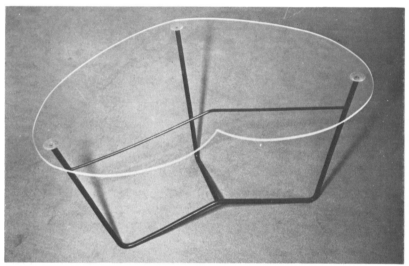

84

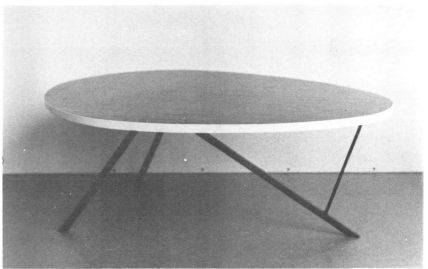

8 Chair design

CHAIRS

Designing a chair presents a number of interesting but difficult design problems. Many solutions are possible involving a wide range of materials, shapes and constructions.

Whatever final design is chosen, it must satisfy certain criteria:

(a) Comfort.
(b) Strength.
(c) Stability.
(d) Good appearance.

The basic shape will be governed usually by its function. A chair for use in a bedroom will be different from one for use in a kitchen.

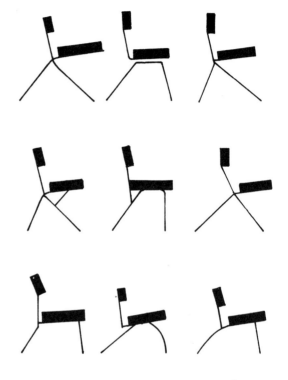

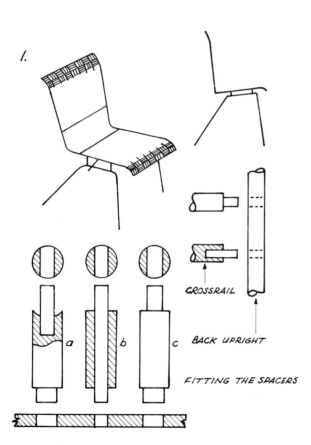

1.

CROSSRAIL

BACK UPRIGHT

FITTING THE SPACERS

CHAIRS 1

MAIN FRAME – BDMS rod 9 mm diameter.
LEGS – BDMS or duralumin strip 16 or 18 mm wide and 6 or 8 mm thick or BDMS rod 9 mm diameter.
BACK and SEAT – Seagrass.

All joints brazed.
Spacers are needed between the main frame and the legs to facilitate the weaving of the seagrass.
Fit the spacers by one of the following methods:

(a) Turn a spigot on the bottom end of the spacers. Drill the top ends of the spacers to take a locating pin and curve the top ends to fit the main frame. Braze the joints.
(b) Drill through the spacers for a locating pin. Curve the tops of the spacers to fit the main frame. Braze the joints.
(c) Turn a spigot on each end of the spacers, rivet them in position and then braze. This method will leave an ugly gap in the joint.

CHAIRS 2

FRAME – Steel tube 16 mm square.

RAILS – BDMS rod 8 mm diameter.

BACK and SEAT – Sponge plastic on plywood or blockboard.

Form the bends at the back by sawing a notch nearly through the tube. Bend to close the gap and braze the joint.

Fix the back as shown in the drawing, by inserting dowels to give grip to the screws. Put an epoxy adhesive on the joint before screwing it together. Upholster the back before or after assembly.

Use one of two methods to position the joints for brazing:

(a) Make a small square plate to fit the inside of the tube. Position this by dropping a rivet into a drilled hole or screw the plate in position with a self-tapping screw, or

(b) Use a short length of tube or steel rod that fits inside the tube to form a dowel. Drill a hole in the other square tube the same size as the dowel.

Assemble the joint. Make up two clamping plates and bolt them across the joint in order to hold the joint tightly together while it is being brazed.

Fix the crossrails to the legs by drilling holes in the legs the same diameter as the crossrails and braze the joints.

For details of the upholstering of the seat, see page 91.

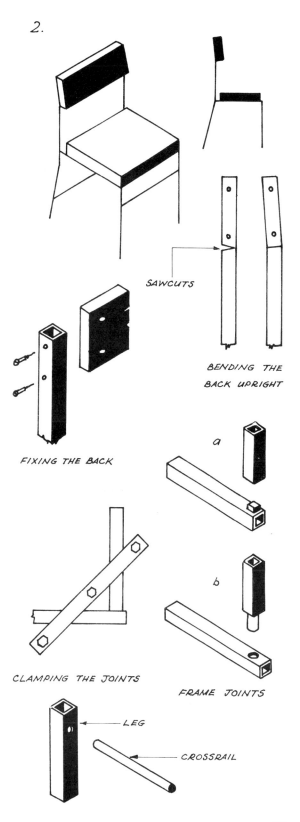

2.

SAWCUTS

BENDING THE BACK UPRIGHT

FIXING THE BACK

a

b

CLAMPING THE JOINTS

FRAME JOINTS

LEG

CROSSRAIL

CHAIRS 3

SIDE FRAMES – Steel tube 16 mm diameter.

BEARERS – Steel tube 12 mm square.

BACK RAIL – BDMS rod 10 mm diameter.

BACK UPRIGHT – BDMS strip 38 mm wide×5 mm thick.

BACK – Three layers of 5 mm plywood glued together and bent to shape and faced with Formica.

SEAT – Sponge plastic on 12 mm plywood.

Jig bend the two side frames first.

Shape the ends of the bearers to fit the round tube and braze in position using a short length of tube or rod as a dowel to locate the joint.

Drill the legs to take the back rail and braze the joints.

Butt braze the back upright to the rail and rivet it to the bearer making use of a BDMS strip.

Form the back from three layers of 5 mm thick plywood, glued and bent round a suitable former. When set, trim to shape and face the front face with Formica.

Use countersunk wood screws to fix the back to the back upright.

Finish all over with an egg-shell black polyurethane paint.

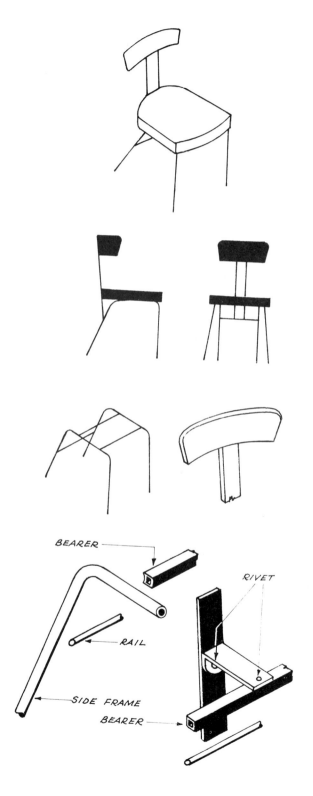

BEARER

RIVET

RAIL

SIDE FRAME

BEARER

CHAIRS 4

SIDE FRAMES – Steel tube 16 mm diameter.
RAILS – BDMS rod 18 mm diameter.
BEARERS – Steel tube 12 mm square.
BACK SUPPORTS – BDMS strip 12 mm×6 mm.
BACK – Three layers of 5 mm thick plywood glued
   and bent on a former. Front of the back faced
   with Formica.

Construction similar to Chair 3 with all joints
brazed.

4.

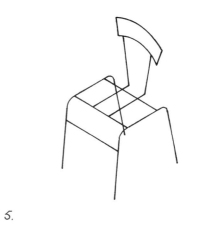

CHAIRS 5

FRAME – Steel tube 16 or 18 mm square.
BACK – Three layers of 4 or 5 mm plywood.
SEAT SUPPORT – Pirelli webbing.

To locate or position the parts of the frame for
brazing, insert a tubular dowel in the end of each
tube and drill a hole in the mating piece the same
size as the overall diameter of the dowel tube. Hold
the joints together with clamping plates and braze.
Glue plug inserts made of plastic, metal or wood in
the open ends of the tubes or, alternatively, plug
the ends with a paste resin.
Make the back from three layers of 4 or 5 mm thick
plywood, glued together and clamped to a suitable
former.
Finish all metal parts with an egg-shell
polyurethane paint.
Use Pirelli webbing or upholstery springs for the
seat. Details of fixing the webbing are given on
page 98.
Make the cushions from thick sponge plastic
covered with material.

5.

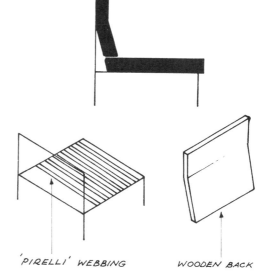

'PIRELLI' WEBBING   WOODEN BACK

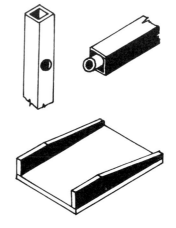

WOODEN FORMER FOR SHAPING THE BACK

EXAMPLES OF CHAIR DESIGN

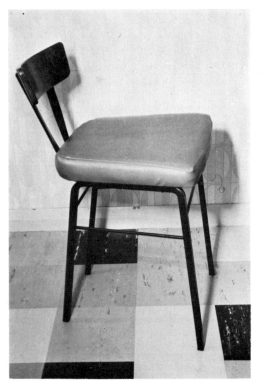

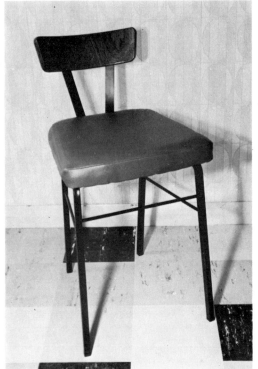

SEATING FOR STOOLS AND CHAIRS

1. G.R.P.

A pattern or mould will be needed on which to form the polyester resin. This can be made from solid wood or can be built up with plywood and wooden battens.

(a) Brush two or more coats of sealer on the mould and on the top face of the baseboard. The quality of finish on the laminate (moulding) will depend on the quality of finish on the mould. Rub down with wet and dry paper and wash down between each coat of sealer.

(b) When a satisfactory finish has been obtained, apply release agent. Remember that if you use a proprietary wax polish it must not contain silicone.

(c) Make the moulding using three or four layers of medium weight mat, resin, pigment and catalyst.

2. Sponge plastic on plywood

Use sponge plastic between 25 and 50 mm thick and cut it larger than the plywood base allowing about a 9 mm overlap all round. Drill the base with 4, 6 or 8 holes about 12 mm diameter so that the air can escape when the plastic is compressed. Glue the plastic to the top face of the base.

If a curved seat is needed, form it from two or three layers of 4 or 5 mm plywood. Glue the pieces of plywood together and either clamp them to the curved top of the stool frame or form the shape over a batten as shown in the drawing.

The covering material can be either a closely woven cloth or vinyl sheet.

If the seat is a flat one, put on the covering material in one piece making a box pleat at the corners or stitching it at the corners with thread.

If the seat is a curved one, make the sides and the top from separate pieces of material and stitch them together.

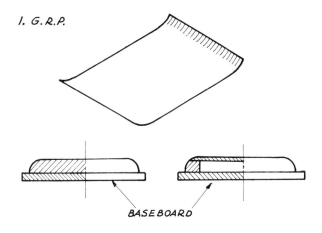

1. G.R.P.

BASEBOARD

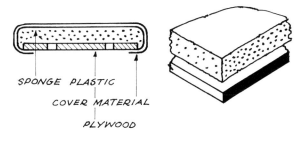

2. SPONGE PLASTIC ON PLYWOOD

SPONGE PLASTIC

COVER MATERIAL

PLYWOOD

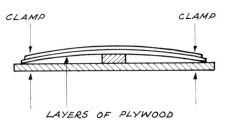

CLAMP   CLAMP

LAYERS OF PLYWOOD

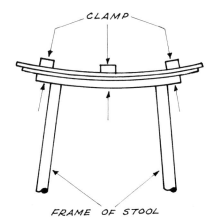

CLAMP

FRAME OF STOOL

9 Solutions to forming and constructional problems

MAKING BENDS IN ROD OR TUBE

Bends can be formed in tube by

(a) using bending springs – these fit inside the tube and a spring is needed for each bore size;
(b) filling the tube with sand or lead – this method is useful only when a few bends have to be made;
(c) using a bending machine – this is the most satisfactory method.

None of the methods will form a bend of small radius without distortion taking place.

Bends in rod can be made when the metal is hot or cold. Bending hot is easier but exact positioning of the bend is more difficult.

Bending jigs

The use of jigs for bending gives the most satisfactory results especially when two or more bends have to be made exactly the same. The jigs can be made from either T iron or 12 or 16 mm square mild steel screwed to a baseplate about 6 mm thick. Drill holes along the length of the baseplate so that studs can be positioned at different centres. Make the studs a push fit in the holes of the base or thread them and fix them to the base with nuts and washers. A length of square mild steel screwed to the underside of the base will enable the jig to be held in the vice more easily.

A strip of mild steel bolted to the baseplate can be fixed in such a way that it can be used to check the angle of the bends.

When attempting to bend heavy metal, always try to make use of lengths of tubing to gain extra leverage.

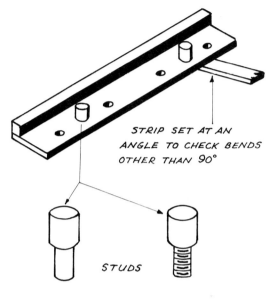

STRIP SET AT AN ANGLE TO CHECK BENDS OTHER THAN 90°

STUDS

PUSH FIT   THREADED

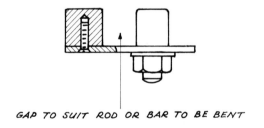

GAP TO SUIT ROD OR BAR TO BE BENT

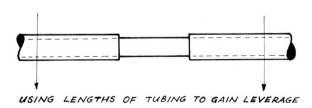

USING LENGTHS OF TUBING TO GAIN LEVERAGE

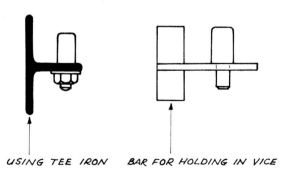

USING TEE IRON   BAR FOR HOLDING IN VICE

Bending Square Steel Tube

1 Bend a piece of 12 or 16 mm square steel tube, heated to a bright red, in the vice. The result is complete collapse of the tube. The metal stretches on the outside of the bend and contracts on the inside of the bend causing considerable distortion.

2 Bend a piece of similar tube, heated to a bright red, in a simple bending jig. The resulting bend will be better than the one done in the vice, but there will still be distortion.

3 Make a similar bend in a jig using a larger bending former 'X'. This bend will show an improvement.

4 Make a trial bend, hot, in a bending jig after having removed metal from the inside of the bend by making saw cuts. When the bend has been made, the saw cuts will need brazing. Therefore, to keep the cuts clean, apply borax before heating to make the bend. First, try six saw cuts about 9 mm apart, sawing through the tube all but the wall thickness.

Distortion will be considerably reduced but a 90° bend will not be possible without the outside and inside faces becoming concave and the side faces bulging. The sawcuts close up on the inside of the bend and open up near the outside of the bend.

5 Experiment with more saw cuts, say 6 mm apart, and this time saw only half-way through the tube.

The quality of the bend will show marked improvement.

6 Increase the number of saw cuts to 12, spaced 3 mm apart.

A 90° bend, with little distortion, is now possible.

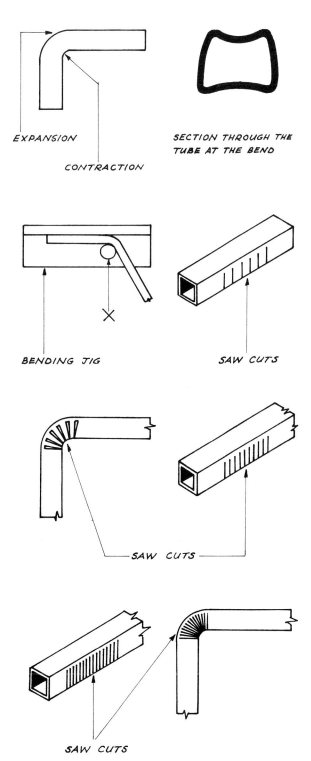

EXPANSION

CONTRACTION

SECTION THROUGH THE TUBE AT THE BEND

BENDING JIG

SAW CUTS

SAW CUTS

SAW CUTS

DESIGN FOR A BENDING JIG

1 BASE – BDMS flat
2 FORMER – BDMS rod
3 BOLT – BDMS
4 ARM – BDMS strip
5 TUBE TO BE BENT
6 GUIDE BAR – Aluminium
7 PRESSURE PLATE – BDMS
8 STOP PIN – BDMS rod

The sizes of the metal for making the bending jig will be dependent on the size and wall thickness of the tube to be bent.

The guide must be made of a soft metal so that the tube will not be marked.

The bolt must be dimensioned so that the former and the arm can swivel when the bolt is tightened.

Square or round tube can be bent by modifying the former and guide bar.

To bend square tube – make the former and guide bar with flat faces.

To bend round tube – make the former and guide bar with concave faces to match the curvature of the tube being bent.

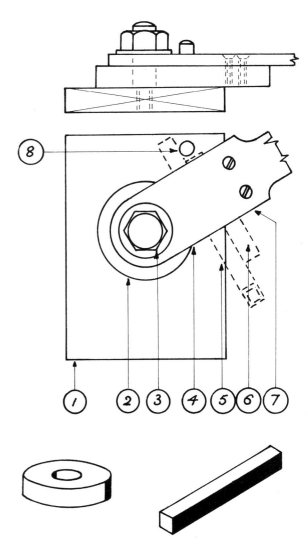

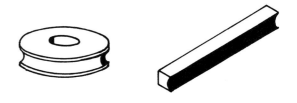

FORMER AND GUIDE FOR SQUARE TUBE

FORMER AND GUIDE FOR ROUND TUBE

94

JOINTING METHODS FOR ROD AND TUBE

1. Rod

(a) Screw together with round-headed or counter-sunk screws. Curve the end of the rail to mate with the curvature of the other member.

(b) Drill a hole in both members. Curve the end of the rail to a fit and insert a pin to hold the two parts together whilst brazing.

(c) Drill a hole in one member and turn a spigot on the end of the rail to fit. Here a gap will be left where the square shoulder meets the curvature of the rod.

(d) Curve the rail to fit the other member and butt braze.

2. Square tube

The main problem when brazing square tube is holding the parts together during the brazing operation.

(a) Shape a piece, say, of 4 or 5 mm thick mild steel to fit the inside of the tube. Drill a hole in the block and the tube. Position the block to the tube by means of a pin that is a push fit in the holes.

(b) Make hardwood plugs to fit in the ends of both pieces of tube and glue them in position with an epoxy adhesive. Joint the two pieces of tube together by means of hardwood dowels in the same way as dowelling a wood joint. Use an epoxy adhesive to fix the joint together.

(c) Obtain a piece of round tube that is a push fit inside the square tube or turn a piece of mild steel rod so that it is a push fit inside the square tube. Drill a hole in the other member to suit the tube or rod.

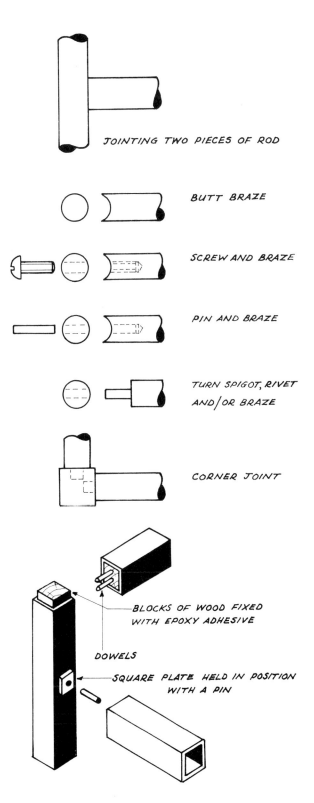

JOINTING TWO PIECES OF ROD

BUTT BRAZE

SCREW AND BRAZE

PIN AND BRAZE

TURN SPIGOT, RIVET AND/OR BRAZE

CORNER JOINT

BLOCKS OF WOOD FIXED WITH EPOXY ADHESIVE

DOWELS

SQUARE PLATE HELD IN POSITION WITH A PIN

3. Rod and tube

(a) Drill a hole through both walls of the tube to suit the diameter of the rod and braze the joint.

(b) Drill a hole through both walls of the tube. Turn a spigot on the end of the rod to fit in the holes and braze the joint.

(c) Drill a hole through both walls of the tube to suit the diameter of the other tube to be brazed to it.

(d) Drill a hole through one wall of the tube to suit the diameter of the other tube and braze the joint.

(e) Obtain or machine a length of mild steel rod so that it will fit inside the tube. Drill a hole in the other tube to fit the diameter of the rod and use the rod as a dowel to hold the two tubes together.

(f) As above, but drill a hole through one wall only of the tube.

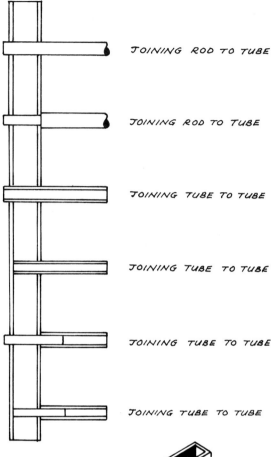

JOINING ROD TO TUBE

JOINING ROD TO TUBE

JOINING TUBE TO TUBE

JOINING TUBE TO TUBE

JOINING TUBE TO TUBE

JOINING TUBE TO TUBE

4. Rectangular tube

Make hardwood inserts to fit the tube. Glue them into the tube with an epoxy adhesive to form tenons. The joint can then be treated as a mortice and tenon. In some instances, parts of the tube will have to be cut away.

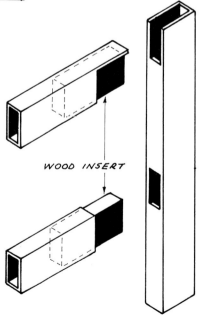

WOOD INSERT

FEET FOR FURNITURE

Protective feet or cups can be obtained made of rubber or plastic. These are stocked by most 'do it yourself' shops but it is much cheaper to buy them in bulk from the manufacturers.

Plastic feet for square tube can be built up from pieces of sheet plastic. It is also possible to make a simple mould and use this to cast the feet. 'Plasticoting' is another alternative.

WOOD FOR TABLE TOPS

Chipboard, plywood, blockboard and laminboard are obtainable in all standard board sizes.

Ranges of thicknesses

Chipboard   – 12–25 mm.
Plywood   –   3–25 mm.
Blockboard – 16–25 mm.
Laminboard – 12–18 mm.

Chipboard provides only an insecure grip for screws. A dowel insert where the screw is to be fixed will increase the grip of the screw. Chipboard, although cheap, has a low strength along its length and horizontal surfaces made of it should be well supported if they have to withstand any weight.

Chipboard has an unattractive surface and for most work the surface will have to be covered. The surfaces of the other boards can be improved too with a covering of Contact, Fablon, vinyl sheeting, Formica etc.

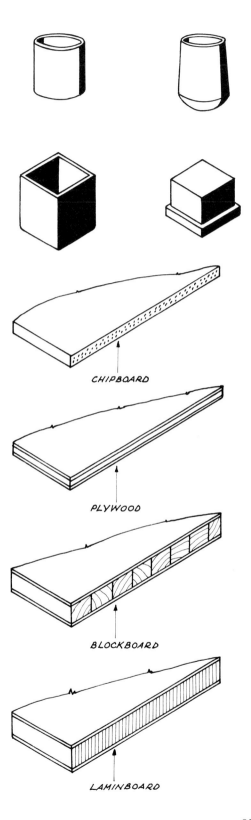

CHIPBOARD

PLYWOOD

BLOCKBOARD

LAMINBOARD

PIRELLI WEBBING

Pirelli webbing consists of a core of rubber between two layers of rayon cords. The webbing is simple to apply, strong and long-lasting. It has a stretch of almost half its original length. To function properly the webbing must be stretched or tensioned when it is applied to the frame.

Attaching to wooden frames

(a) Tack one end of the webbing to the seat frame member making sure that the tacks are at least 6 mm from the end of the webbing. Use 16 mm upholstery tacks or 12 mm clout nails.

(b) Measure from the tacks the length of webbing required to stretch across the frame allowing for the initial tension and mark the webbing.

(c) Stretch the webbing until the mark is in the centre of the opposite seat frame member, then tack in position.

(d) Cut off the surplus webbing at least 6 mm from the tacks.

Make sure the tacks or nails are vertical. If they are driven in at an angle, the edge of the head may cut the rayon cords and cause a fracture. Round off the inside edges of the frame; a sharp edge will cause undue wear on the webbing. Patented clips are available for attaching the webbing to the frame. The clips are squeezed onto the end of the webbing in the vice, then placed in a groove or mortice cut 4 mm wide, 14 mm deep and angled inwards at 10° to 15°.

Attaching to metal frames

Wire clips are available for 38 mm and 50 mm webbing widths only.

The clips are fixed to the webbing by staples and plates.

(a) Punch holes in the webbing as shown.

(b) Fix the clips to the webbing by means of the staple and backing plate.

(c) Drill 2.5 mm holes in the tube or angle iron of the framing to take the clips. When drilling the metal, make sure the holes are at top dead centre or the prongs may disengage from the holes.

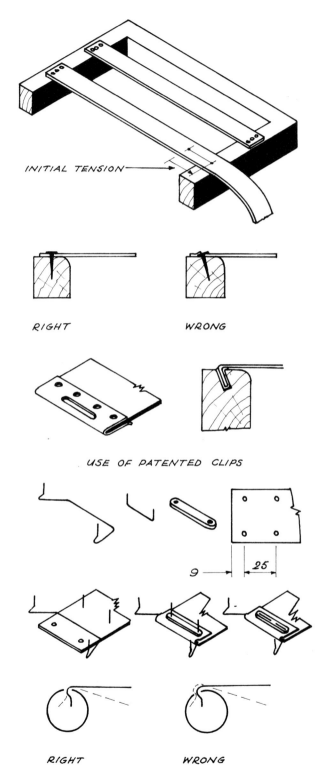

INITIAL TENSION

RIGHT   WRONG

USE OF PATENTED CLIPS

9   25

RIGHT   WRONG

SEAGRASS AND SEATING CORD

Traditional methods of weaving with seagrass or seating cord are usually confined to weaving in a double layer across the seat rails. Rails on metal furniture are more slender than their wood counterparts and it is possible on metal furniture to weave in a single or a double layer, the former usually being the neater of the two.

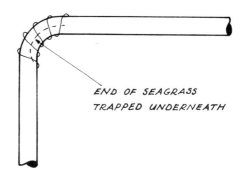

END OF SEAGRASS
TRAPPED UNDERNEATH

(a) Trap about a 50 mm length of seagrass near one corner of the frame. If the seagrass tends to slip bind it with Sellotape to hold it in position.

(b) Work out the pattern, i.e. how many strands are going to form the weave, 2, 3, 4 or 5. A space must be left between each group of strands to allow for cross weaving. If the pattern is being formed with two strands then three strands must be wound round the frame for cross weaving. If a pattern of three strands is being used then wind round four strands, etc. Wind one more strand around the frame than you are using for the pattern.
This first weave of the pattern must be fairly slack because of having to cross weave through it.

(c) When the end of the frame has been reached, continue around the corner and cross weave.

(d) When knotting on a fresh length of seagrass, arrange for the knot to be underneath where it will show least. On a chair back a knot in the front or in the back will be unsightly so arrange for the join to come alongside the metal frame and trap the seagrass in much the same way as when starting.

(e) Finish by knotting underneath or by weaving the last short length through the pattern where it will show least.

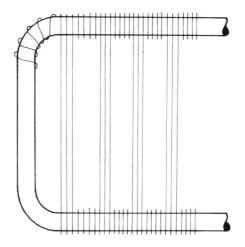

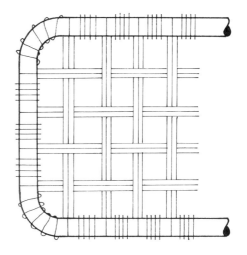

A HOT WIRE CUTTER to cut expanded polystyrene up to 120 mm thick

The box

Make this from 16 or 18 mm blockboard, glued and screwed together.

The frame

Use BDMS or duralumin strip 16 mm wide and 4 mm thick. Bend this on edge at 90°.

The wire

Nichrome wire about 4Ω is suitable. Tension the wire by means of a coil spring. Fit it to the base by means of a small bracket and fit it to the frame with the spring and a brass strip.

Transformer

The transformer, reducing the voltage to 6 or 12 volts, must be capable of taking up to 48 watts. If a bulb is not included in the circuit, an output of 6 volts is sufficient.

Circuit

Wire a switch into the live lead from the plug. Connect an insulated lead from the 12-volt transformer output to the base of the metal frame. Wire the other output from the transformer via a 12-volt 24-watt bulb to the base bracket. The degree of heat that develops in the wire can be varied by using car bulbs of different wattage. The temperature to aim at is between 200 and 250 °C. Avoid getting the wire red hot, because at that temperature the polystyrene will give off fumes which are pungent and irritating.

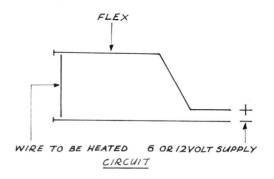

FLEX

WIRE TO BE HEATED   6 OR 12VOLT SUPPLY
CIRCUIT

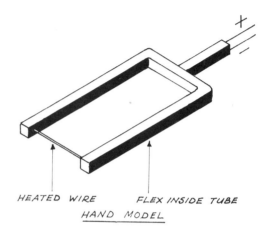

HEATED WIRE   FLEX INSIDE TUBE
HAND MODEL

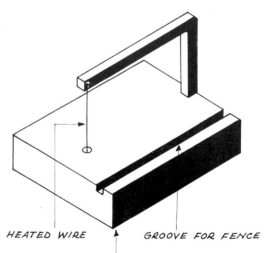

HEATED WIRE   GROOVE FOR FENCE

CASE TO CONTAIN THE TRANSFORMER
AND A BULB IF A 12 VOLT CIRCUIT IS
BEING USED
TABLE MODEL

POLYSTYRENE CUTTER

Expanded polystyrene is most easily cut with a hot wire. The heat to the wire is provided by a low voltage electrical supply of 6 or 12 volts.
A table model is the best option when large pieces of the material have to be cut, but for cutting small pieces a hand model is the best.

POINTS TO CONSIDER

1 Type of wire to use, i.e. what metal?
2 What diameter of wire?
3 What is the best voltage to use?
4 The wire must be held taught for efficient cutting. How can this be done?
5 A table model will be more versatile if it is capable of cutting at an angle as well as square. How can this be done? Will it be easier to tilt the wire or tilt the table?
6 It would be an advantage if there were a simple fence to ensure that a parallel cut is being made.

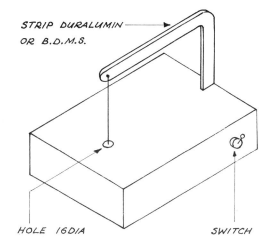

STRIP DURALUMIN OR B.D.M.S.

HOLE 16 DIA   SWITCH

SWITCH

12V

BULB   4 Ω HEAT WIRE

TRANSFORMER

13 AMP PLUG   EARTH

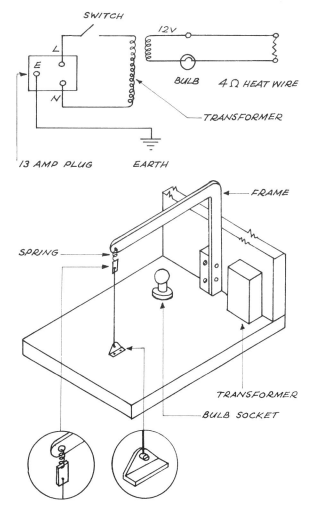

FRAME

SPRING

TRANSFORMER

BULB SOCKET

10 Photographs of solutions to various design problems

Collapsible canoe trailer

Candle holders made from BDMS sheet, BDMS tube. Joints soft soldered

Right Two designs for chessmen made from duralumin and plastic rod

Chessmen in walnut and duralumin, using epoxy adhesive to bond them together

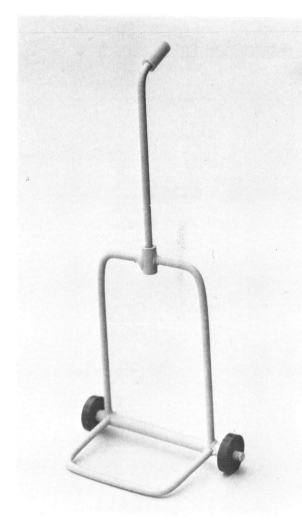

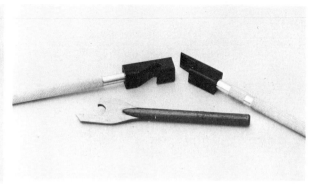

Bottle openers
Top. Square tube and duralumin
Bottom. BDMS strip and rod. Joint brazed

Drilling machine – not completed

Mock-up for a shopping trolley

Pepper pot in duralumin tube and rosewood.
Salt container machined from duralumin rod

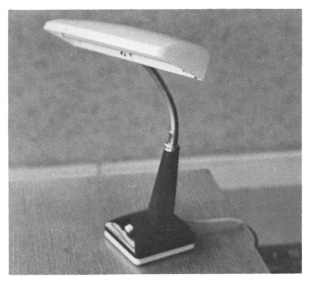

Modern manufactured strip desk lamp

Wall light in duralumin

Adjustable lamp made in glass-reinforced polyester and aluminium alloy

Standard lamp
Base – hardwood
Stem and fittings – aluminium alloy
Shades – G.R.P.

Bench shears

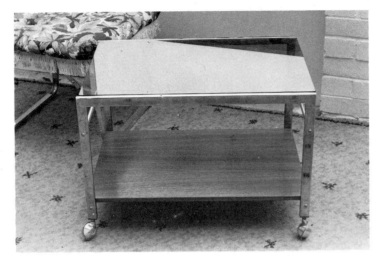

Occasional table
Top – glass
Main frame – rectangular steel
tube, joints brazed
Bottom shelf – chipboard faced
with laminate

Bed table

Small table
Top frame – steel angle
Top – pattern embedded in resin

Abstracts in wood

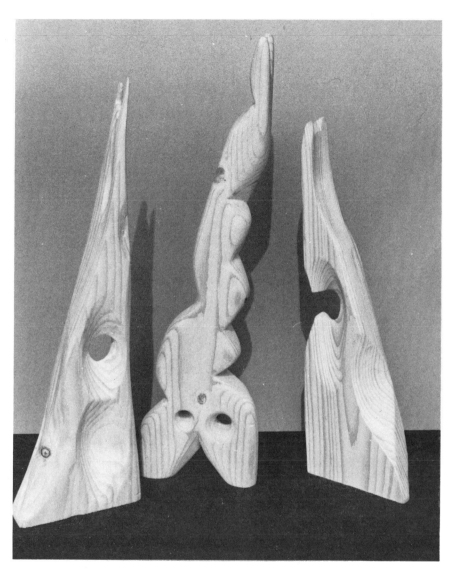

Abstracts in plastics

107

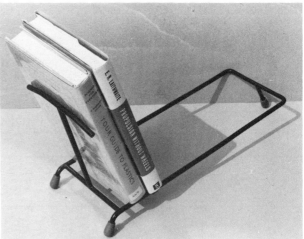
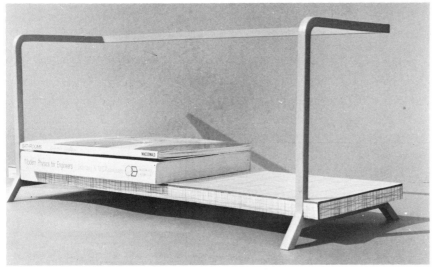

Book racks
Frames – BDMS
Shelves – wood faced with laminate

Experimental radio-controlled truck

Tractor (less engine) for use on a small holding

Go Kart – experimental model

Centre lathe constructed from BDMS flats bolted together

Simple teaching machine
Case made from acrylic sheet

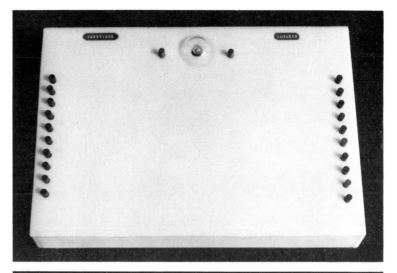

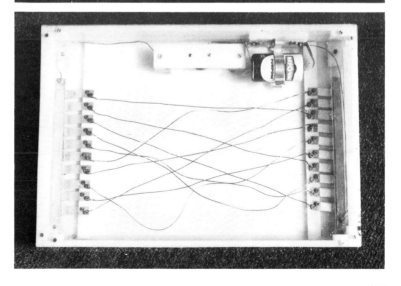

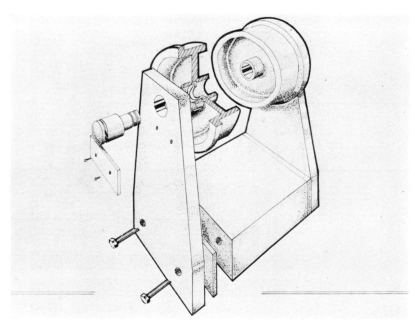

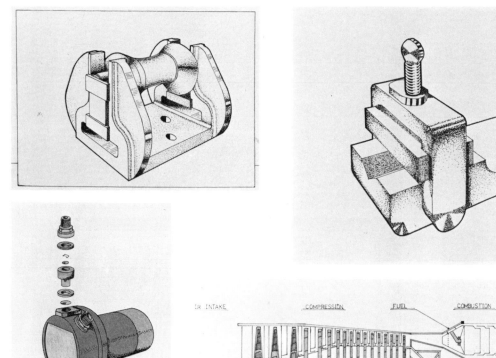

AIR INTAKE   COMPRESSION   FUEL   COMBUSTION   EXHAUST

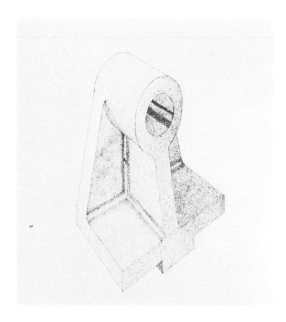

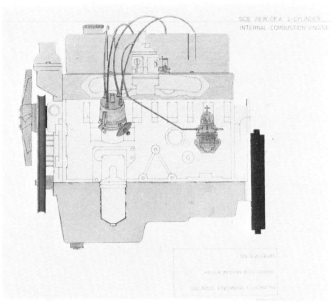

IAN BLACKBURN

HELENA MODERN HIGH SCHOOL

CASE MODEL 3 TECHNICAL ILLUSTRATION

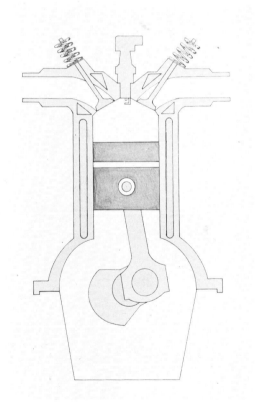

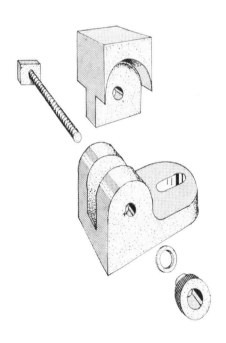

113

Telephone shelf with
a plate glass top

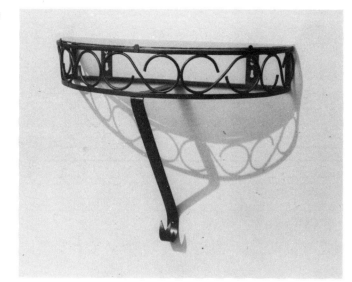

Lantern in brass with
a BDMS bracket

Lantern made from
BDMS sheet

Battery clock made from acrylic sheet

Pendulum battery clock in sheet acrylic. Made by Presto, Japan

Wall light made from chipped block glass set in resin.
The back is reinforced with one layer of glass mat.
Illumination supplied by a 300 mm-long strip light

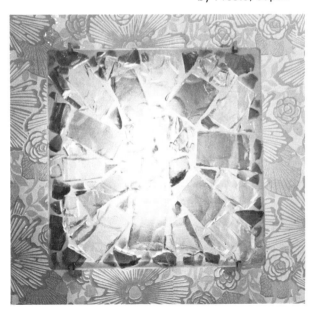

115

Project Technology. Gas-fired furnace

Project Technology. Stand for a portable drill

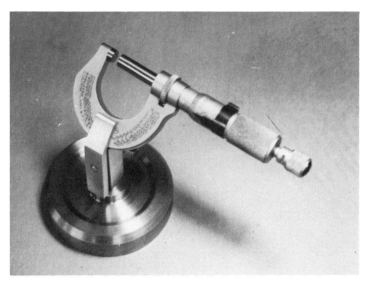

Project Technology.
Micrometer stand

Table barbecue

Ashtray

Simple hearth
set

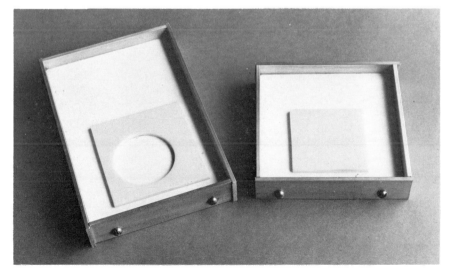

Mould from acrylic sheet
to be used for casting
small resin wall panels

11 Sample problems in design and construction

1a Show diagrammatically how rotary motion (motion in a circle) can be changed into linear motion (motion backwards and forwards in a straight line).

 b Give examples of the use of this principle in everyday life.

 c Design a toy or simple mechanism which uses this principle.

2 The front elevation (front view) of a cone is exactly the same shape as the end elevation (end

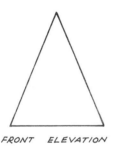

FRONT ELEVATION   END ELEVATION

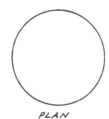

PLAN

view). The true shape is evident only when the plan is drawn in the shape of a circle. Make sketches of four other shapes that satisfy the same conditions.

3 Design a fitting to hold a push bike upright whilst working on it.

4 A clock face must be both clear and easy to read. Sketch two clock faces that satisfy these conditions and two that do not.

5 Explain, with sketches, how you would make a triangular assembly puzzle using acrylic sheet.

6 Symbols are a form of language which gives information in pictorial form.

 a Draw four road sign symbols.

 b Make a drawing of an identifying plaque for your own bedroom.

 c Make drawings of symbols that could be fitted to four different classrooms to indicate the subject that is taken in the rooms. Assume that the symbols will be made from acrylic sheet.

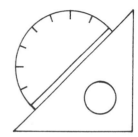

7 Draw an interesting shape such that four of them when fitted together will form a square.

8 Draw four shapes of fishes that could be used to make a brooch out of thin copper sheet.

9 Make sketches of an article that could be produced in numbers to sell at a school fête. Two points are important:

 a The design must be easy and cheap to make.

 b It must sell well to make a profit.

10 A simple electric lamp is needed to serve as a night light in a tent. It is intended that the lamp will run off a 12-volt car battery and, when in use, will be fixed to a tent pole. Make drawings of a lamp that you think you could make.

11 Draw two rectangles 120×75 and **(a)** divide one rectangle into an interesting pattern using straight lines only and **(b)** divide the other into an interesting pattern using straight and curved lines. Colour the shapes.

12 By means of simple line diagrams, express the following situations:
explosion, movement, light, growth, heat, collision.

13 Using simple line diagrams only (labelled if necessary) and with no written explanation show how to:

  a check the squareness of two adjacent edges of a piece of metal or wood;
  b fix a drill in a chuck;
  c mark a centreline on a piece of metal with oddleg calipers;
  d draw a 50 mm radius circle with a compass.

14 Broken hacksaw blades make good marking knives but are difficult to hold Design a handle that could be made to hold these blades efficiently.

15 Given a length of 6 mm square rubber, design a simple shape that can be propelled by the rubber in its tensioned state.

16 Design a sculptured form to represent 'buildings' using lengths of steel tubing.

17 Using two lengths of wire, make a three-dimensional shape which gives the impression of speed.

18 Design a barrow or a truck suitable for a child of four years of age, which could be made in the school workshop.

19 Make drawings of a night light suitable for a child's bedroom. The wattage of the bulb is 15.

20 A child's rattle is to be made in the school workshop. List the priorities, e.g. materials must be non-toxic, and make drawings of one that you could make in the school workshops.

21 A standard lamp is to be made from a length of steel tube 25 mm diameter and with a wooden base 200 mm diameter and 25 mm thick. How would the tube be fitted to the base?

22 Provide a solution to the same problem if the tubular stem was made of duralumin.

23 A bulb holder, which is to be fitted to the top of the stem of a standard lamp, is threaded internally with a ½ inch×26 t.p.i. thread. How can this be done if the stem is made of steel tube and aluminium alloy tube?

24 Show, diagrammatically, how to wire a 13 amp plug.

25 If you had to design and make a salt cellar out of metal, give details of the kind of filler plug you would make.

26 How would you fit a 6 mm plate glass table top to both a wooden and metal underframing?

27 If the top were made of Perspex, what fixing method would you use?

28 A table lamp is made of a metal tubular stem and a metal base. A press button switch is to be fitted to the base. Make a diagram of the wiring, using a 13 amp plug.

29 An egg-timer glass has to be fitted to a wooden backplate in such a way that it can be rotated. How can this be done?

30 Four rectangular pieces of Perspex are fixed together in an open box shape to form a lamp. Explain how you would make provision for holding the bulb.

31 A table top is made of blockboard faced with Formica. Sketch three forms of edging that could be used to hide the edge of the blockboard.

32 Describe how you would make the joint at the corner of a square frame made out of 16 mm square steel tube. What form of clamp would you use to hold the joint together whilst it is being brazed?

33 The stem of a reading lamp is made of 16 mm O.D. aluminium alloy tube. An arm of 12 mm O.D. alloy tube must be capable of being locked at any angle to the stem at 'A'. Design a clamping method capable of doing this.

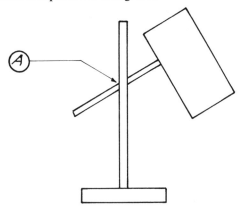

34 A rectangular table top is to be faced with ceramic tiles. The top is made of 12 mm thick chipboard edged with a frame made of steel angle. How would you construct the frame and fix the chipboard to it?

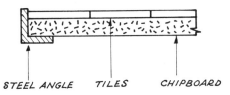

STEEL ANGLE   TILES   CHIPBOARD

35 A stool has four splayed tubular steel legs and a wooden top. What method of construction would you use to join the top to the legs?

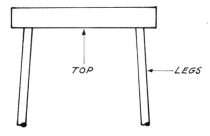

TOP   LEGS

36 A collet screwdriver has blades made with two projecting lugs to prevent rotation. How can this be done?

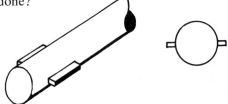

37 A chessman is to be made from a length of plastic rod with a knurled duralumin base fitted to it. Describe two methods of construction.

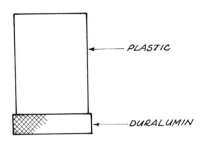

PLASTIC

DURALUMIN

38 A small table has a top made of 6 mm thick Perspex sheet. The edges are to be curved as shown in the drawing. Design a tool that can be used to radius the edges instead of having to file them.

39 You have to make four steel washers, 16 mm diameter and 1 mm thick with a 6 mm clearance hole in the centre, from BDMS sheet. Design a fitting to hold the washers for machining.

40 The sketch shows the end frame of a table unit. The main frame is made of 16 mm×18 s.w.g. square steel tube and the crossrail is made of 12 mm diameter×18 s.w.g. round steel tube. Give details of the construction that could be used for the joints at 'A' and 'B'.

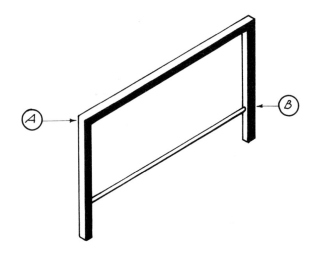

41 Give details of how you would overcome the following joint problems:

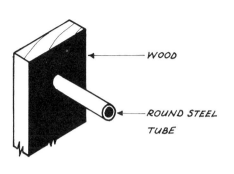

WOOD

ROUND STEEL TUBE

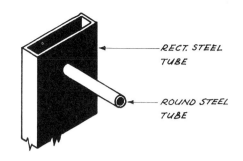

RECT. STEEL TUBE

ROUND STEEL TUBE

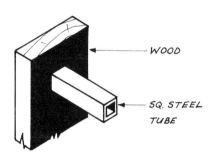

WOOD

SQ. STEEL TUBE

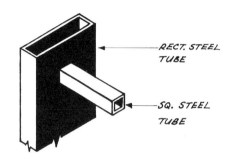

RECT. STEEL TUBE

SQ. STEEL TUBE

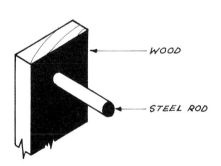

WOOD

STEEL ROD

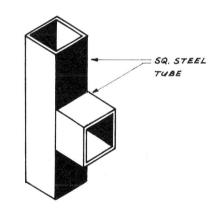

SQ. STEEL TUBE

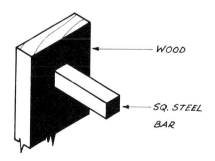

WOOD

SQ. STEEL BAR

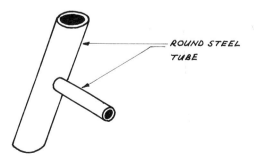

ROUND STEEL TUBE

121

42 How would you make a plug or an end cap for both square and round tube?

43 The drawing shows part of the end frame of a kitchen stool with a curved upholstered top. How can the tubular steel legs be joined to the curved steel plates?

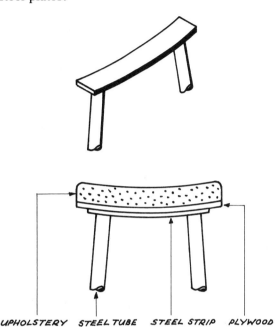

UPHOLSTERY   STEEL TUBE   STEEL STRIP   PLYWOOD

44 A G-cramp will be more efficient if a collar is fitted to the end of the screw in such a way that it can rotate and swivel at any angle. How can this be done?

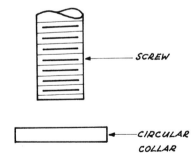

SCREW

CIRCULAR COLLAR

45 A mild steel bar 200 mm long and 40 mm diameter has to be machined as shown. Explain how this can be accomplished on a centre lathe which has a headstock bore of 16 mm diameter.

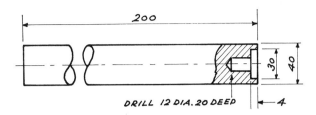

200

30   40

DRILL 12 DIA. 20 DEEP   4

46 Design and make a puzzle that would be suitable for blind people.

47 Design a kinetic sculpture that can be operated by the simplest natural or mechanical means.

48 Design a toy suitable for use on a long train journey.

49 Design a toy clock that is suitable for teaching a child to tell the time.

50 Analyse the problems involved in designing a stool suitable for use at a bar in a hotel. Show clearly the ergonomic factors that influence the design. Consider alternative structures and materials that could be used. Give reasons for your final choice of design and materials.

51 A small model-maker's vice is required to the following specifications:
   a The jaws to be 25 mm wide and 10 mm deep. There need be no 'cut away' below the jaws which should be smooth-faced and need not be hardened.
   b The jaws must move with a parallel motion.
   c The only machine tool which can be used is a pedestal drill but all hand tools are available.
   d The vice is to be bolted to the top of a bench by one 10 mm bolt so that the whole unit can be swivelled.
   Design a vice to the above specification making working drawings of each part, stating the material involved. A set of instructions for making the parts should accompany the drawings.

52 Taking into account shape, texture and colour, make an abstract wall plaque suitable for the wall of a lounge.

53 Design a fitting to hold a drink container on the front of a bicycle.

54 Design a maze using colour to confuse the issue.

55 The T-slots on a shaping machine have become clogged with swarf. Draw a tool that would be suitable for removing the swarf.

56 An electronic engineer has a lot of small components to inspect. He wishes to utilize a magnifying glass 50 mm diameter which focuses clearly on an object 75 mm away from the glass. Sketch annotated ideas for a suitable free-standing holder for the magnifying glass which will enable him to view the components. Suitable dimensions of parts should be indicated.

57 The internal threads of a lathe chuck should be cleared of chippings before the chuck is screwed to the lathe. Make a tool that will do this efficiently.

58 The illustration shows a screw that will fit the threaded socket in the base of a camera. Design a universal clamp that can be attached to the camera by means of this screw. The clamp should enable the camera to be clamped in any conceivable position to such things as table tops, shelves etc., while time exposures or delayed action photographs are being taken.

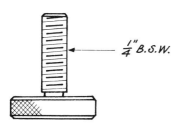

59 Your father requires a house number plate which is to be mounted on the wall at the side of the door. Design and make one in any suitable material or combination of materials.

60 A fibreglass shade has been purchased with the sizes shown in the drawing. Design a fitting to take the shade.

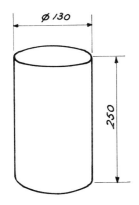

61 Design a clip that will hold a tie neatly to a shirt front.

62 A decorative porcelain tile 108 mm square and 5 mm thick is to be used as a cutting surface for a cheese board. Design a cheese board that will accommodate this tile and also hold a cheese knife.

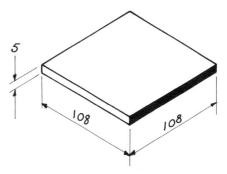

63 A set of three ball-point pens are used on a writing table. Design a block to hold them using any suitable material.

64 Design a toy fitted with four wheels so that it can be pushed or pulled along and is suitable for a young child.

65 Details of a table lighter mechanism are given. Design a table stand or case that can be made from wood, plastic or metal or combinations of these materials.

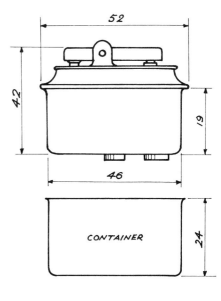

66 Design a small sandwich tray that can be made out of wood, metal or G.R.P.

67 A container for matches is required that will fit on a wall in a kitchen. Provision must be made so that the matches may be struck when the container is fixed to the wall and when the container is detached from the wall.

68 Design a simple fitting or attachment that can be fitted to a TV set or to a TV table suitable for holding a Radio Times or TV Times.

69 Design a swivel head for a camera tripod. Standard thread mounting on the camera is $\frac{1}{4}$ inch BSW.

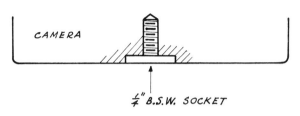

CAMERA

$\frac{1}{4}''$ B.S.W. SOCKET

70 A Harrison centre lathe has an all-geared head, the top of which is fitted with a small cover plate and two gear levers to the dimensions shown. It is difficult to position a sheet drawing or a book when working from them. Make a fitting that will overcome this problem.

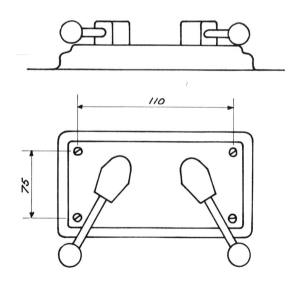

110

75

71 Design a holder for four pints of milk that can be placed outside the door of your house for the daily delivery of milk. Means must be provided for indicating to your milkman the number of pints needed.

72 The machining of semi-circular curves and round knobs is an operation that causes some problems. Design an attachment that will fix either in the toolpost or in place of it that would eliminate these problems.

73 The woodwork room requires a jig to enable dowelled joints to be made quickly and efficiently. Design a jig that can be used for 6 and 8 mm diameter dowels.

74 Short lengths of dowel can be made with a dowel plate. This is a heavy plate of mild steel drilled with a series of holes. The dowel is cut roughly to shape and driven through the plate. In use the plate is held in a metalworker's vice. It is an advantage if a groove is formed along the length of the dowel for the escape of air when the joint is being made. Design a plate that will enable you to shape 6, 8, 10 and 12 mm diameter dowels.

75 The heat from a gas-air blowpipe makes it difficult to braze or silver solder joints using short lengths of solder. Design a fitting that will hold short lengths of solder so that the hand is at a comfortable distance from the flame when soldering.

76 Design a dispenser for a roll of Sellotape 25 mm wide. It can be made to fit on a wall or rest on a table.

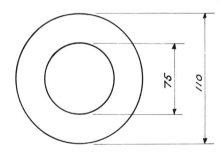

75

110

77 Make a mounting in wood, metal or plastic for a spirit level. Dimensions of the glass holding the spirit are given.

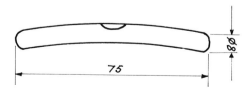

80

75

78 Make a cruet (condiment) set out of metal or plastic tubing and design a stand to hold the set.

79 Design and make a device to convert rotary motion into reciprocating motion. The device is to be hand-operated. The following restrictions are imposed – the base form is to be within 1000 mm and the materials available are plywood, dowel rod, sheetmetal and welding rod.

80 Make a scale model of a chair suitable for occasional use in a lounge. Materials available are metal tubing, glass reinforced resin, sponge plastic and flexible vinyl sheeting.

12 Examination questions, CSE and O level

1 A hull is required for a model boat which is to be radio-controlled. It should be thin, strong, watertight and capable of receiving later fittings relevant to the complete design.

Either **(a)** with the aid of sketches, describe how you would make such a hull, giving full details of the materials, adhesives and fixing devices involved.

Or **(b)** by means of circuit diagrams, etc., describe a simple single channel receiver that you would make for use in such a model. (*Design and Technology UL.*)

2 A design engineer has the problem of coupling two rotating shafts under three different conditions. The first **(a)** is when the two shafts are in line with a small gap between them. The second **(b)** is when the shafts are in line but at right angles to each other and the third **(c)** when the shafts are running parallel to each other.

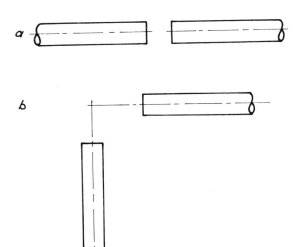

Using simple diagrammatic design sketches, accompanied by brief written explanations, show how you would attempt to solve the problems. (*Engineering Design UC.*)

3 A wrought iron gate requires a means of holding it in the open position. It swings back against a brick wall to which the back plate of the fastener could be attached. It would be an advantage if the gate would fasten itself when swung open against the fastening contrivance. Make a clear sketch of the fastening that you would make which would be fastened to the wall. Explain how you would make it and state the material you would use. (*CSE SREB.*)

4 The centre-piece of a dining table is to be an artistic stand holding three 6-inch tapered candles. Make three small sketches of possible types and from them produce a sketch of the stand approximately full size. The details of the sizes and construction are to be included. (*CSE SREB.*)

5 Design a small dish or tray that may be used for holding sweets or peanuts at parties. Include in the final sketch details of sizes, metal, finish and fastenings, if any. (*CSE SREB.*)

6 Design a teapot stand to measure about $4\frac{1}{2}$ inches across. In your detailed sketch include the metal and finish recommended. Mention any special considerations that you have to bear in mind in the design. (*CSE SREB.*)

7 Design a portable rack or stand that will hold an instrument type of electric soldering iron ready for use. Illustrations of typical soldering irons are shown. Make a list of the points that should be taken into consideration and in your final drawing or sketch include essential details. (*CSE SREB.*)

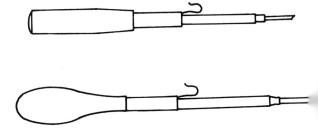

8 Design a shopping trolley that will take a large shopping bag measuring 355 mm wide by 400 mm long by 160 mm deep. Give details of the fixings and the materials required including wheel size. (*CSE SREB.*)

9 A box containing portions of processed cheese measures 110 mm diameter and is 16 mm deep. Design and carefully sketch a stand to hold this. It is to be easily kept clean and readily lifted. Metal, sizes, fixings and other details should be included. (*CSE SREB.*)

10 A door chain is often fitted inside a front door to allow it to be opened part-way with little risk of anyone being able to force an entrance. Design and sketch one that will permit a door to open 130 mm. Include full constructional details. (*CSE SREB.*)

11 Design a portable holder for the electrician's soldering iron shown. The holder must have provision for holding a reel of solder, also shown, so that it can be used from the holder. The holder must ensure that the hot bit will not be able to touch the holder, lead of the iron or the reel of solder. (*CSE EAEB.*)

12 Design a stationery rack to stand on the desk in a headmaster's study to hold notepaper which measures 200×125 mm and envelopes which measure 150×85 mm and provide accommodation for a ball-point pen, a pencil and a letter opener. State the material and the finish you would use. (*CSE SREB.*)

13 The figure gives the external dimensions of the outer case that encloses the battery and the mechanism of an electric clock movement that is suitable for a kitchen clock. It is secured to the clock face by a knurled nut A and a locknut B. They can be made to fit any thickness up to $\frac{1}{2}$ inch by adjusting the lock nut. A brass hanging plate C is fixed to the back of the case but it can be removed if it is not required. The larger hand sweeps a 6 inch circle. You are not required to make or fit the hands.

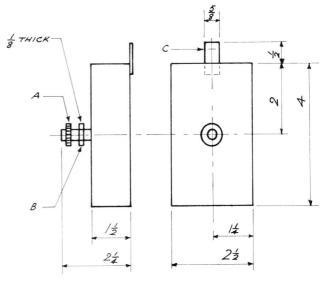

ALL DIMENSIONS IN INCHES

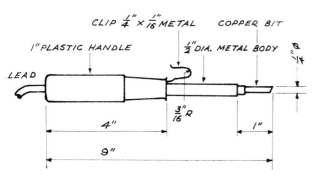

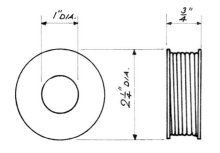

Either (**a**) make a clock-case and face for the mechanism so that it can be used in a kitchen as a wall clock.

Or (**b**) make a clock-case and face for the mechanism so that it can be used as a free-standing kitchen clock that could be placed on a working surface. (*CSE EAEB.*)

127

14 Design a clamp that will hold sheet metal down to a drilling machine table (shown in section) so that the metal can be drilled safely. The clamp must be able to be fixed to the table temporarily (no holes are to be drilled in the table) and then the drill be operated with one hand. (*CSE EAEB.*)

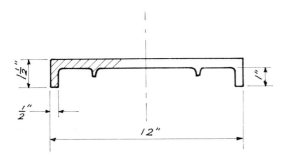

15 A small tea towel airer with two or three horizontal swinging arms, pivoted at one end to a backplate, is required for a kitchen. The arms are to be about 250 mm in length and made from 5 mm diameter mild steel and finally covered with thin-walled plastic tubing, which is readily available. In being asked to design the completed towel airer you should, firstly, consider:

The method you would employ to fasten the rods to the backplate so they can freely swing through 180°.
The shape of the backplate and its method of fastening to the wall.
Suitable materials from which the completed design could be made, assuming a well-equipped school workshop is available complete with a foundry bay.

Now answer:

(a) Make sketches of your completed design, clearly showing the pivoting mechanism and giving the main dimensions. Number or name the separate parts.
(b) Where metal is joined, label the appropriate places on your sketch with the name, or type, of joint employed. Where the metal is shaped, label with the process used.
(c) Make a list, in table form, of all the parts of the design, including the swinging arms, under the following headings:

Part No. or Name	Number Required	Sizes	Material	Finish

(d) Explain, in detail, using sketches as well, if you wish, how you would make one of the joints connected with the pivoting mechanism. Name all the tools and materials used.
(e) Say how you would clean up the completed job and what finish you would apply to the parts, other than the arms. (*CSE SEREB.*)

16 You have been asked to make in the school workshops an extending lamp, on the lines suggested by the sketch.

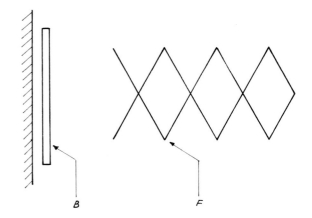

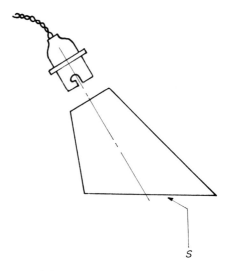

The essential parts are: a backplate B, a frame F and a conical shade S.

The following points should receive your consideration:

The shape of the backplate.

The method of attachment of the frame to the backplate.

The method of attachment of the frame to the shade.

The method of attachment of the backplate to the wall.

Now answer:

(a) Show, by means of sketches, how you arrive at a finished design, adding marginal notes where necessary.

(b) On your final sketch, or sketches, suggest suitable materials for the various parts and add the main dimensions.

(c) Describe, in detail, with sketches, the process of loose riveting the frame members and also the method you would adopt to ensure that the rivet holes are drilled at correct intervals. Give a full specification of the type of rivet you would use.

(d) Show, with the aid of sketches and a written explanation, how the conical shade will be brought to shape and describe, in detail, how the seam joint is to be made.

(e) Indicate the finish you would apply to the three main parts of your design. (*CSE SEREB.*)

17 The painting of external window frames, doors and guttering of a house and garage is frequently done by the 'Do It Yourself' craftsman. This often involves using a ladder to reach work at first floor roof level. One of the problems presented is the need to have both hands free to use the paint brush and to steady oneself on the ladder.

Fig. 1 shows part of a ladder.

Fig. 2 shows two sizes of paint brushes.

Fig. 3 shows a dusting brush.

Design a ladder tray or platform to accommodate a paint tin of 115 mm maximum diameter, 135 mm high and weighing 2 kg, together with one brush of each size illustrated in Figs. 2 and 3.

It is essential that the ladder tray/platform:

(a) fits firmly to any rung of the ladder and can be quickly attached and adjusted to a level position;

(b) when in the working position, does not impede the painter;

(c) when not in use, is easily stored away taking up as little space as possible.

Make complete working drawings, and a list of materials required. Indicate the 'finish' of your choice. All preliminary sketches leading up to your final design, together with any explanatory notes, must be attached to your drawing. Marks will be awarded for the development of ideas in sketch form. Candidates are expected to show by their solutions to this problem an appreciation of the materials, processes and constructions of the craft. (*AEB Craftwork – Metal 'O'.*)

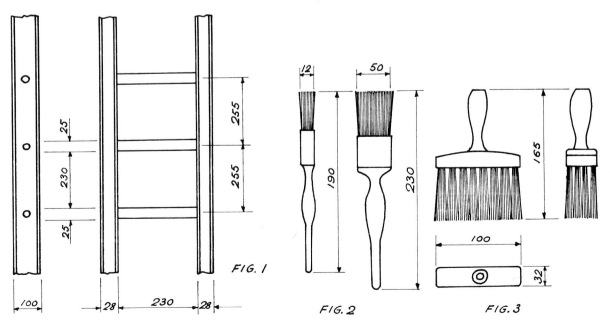

FIG. 1

FIG. 2

FIG. 3

18 One method of jointing pieces of wood together is by means of dowels. In frame construction such as stools, this entails the cutting of several pieces of dowel of the same diameter and short length. When using dowel for joints it is necessary to cut them accurately to length and to groove the small pieces of wood so that the surplus glue is not trapped in the joint.

The diagram illustrates the groove, which is often made by using a dovetail or tenon saw.

Difficulty is experienced in holding small pieces of dowel for cutting and grooving.

Design a wooden jig which will hold dowel rods varying in size from 6 mm to 12 mm in diameter and can be used for cutting:

(a) pieces of dowel with a range from 25 mm to 75 mm in length;

(b) a groove with a tenon saw along the length of the dowel as shown in the diagram.

It is essential that the jig:

can be held in the vice or against the edge of the bench;

includes an easy and accurate adjustable fence or stop permitting the quick cutting of dowels;

ensures that the circular sectioned pieces of dowel are held firmly for grooving.

Make complete working drawings, a cutting list of timber and a list of any fittings and fastenings required.

Indicate the timber 'finish' of your choice.

All preliminary sketches leading up to your final design, together with any explanatory notes, must be attached to your drawing. Marks will be awarded for the development ideas in sketch form. (*AEB Craftwork – Wood 'O'.*)

19 Fig. 1 shows a sectional view of the construction of the window sills in a school corridor.

Design a container for plant pots to be used on the window sills which would satisfy the following conditions.

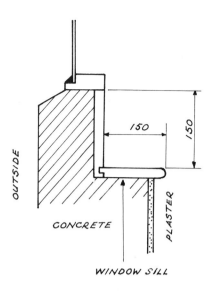

(a) Hold three pots each 120 mm diameter by 120 high.

(b) Permit the watering of the plants without the water draining onto the sill.

(c) Not protrude into the corridor. (*AEB Craftwork – Design 'O'.*)

20 Polystyrene sheet is cut by using an electrically heated wire which is fixed at right angles to a baseboard made from 18 mm blockboard as shown in the diagram.

The art department in a school wishes to cut a quantity of discs, without a centre hole, ranging from 75 mm to 150 mm diameter from square blanks of expanded polystyrene varying from 6 mm to 25 mm in thickness.

Design a jig which can be attached firmly to this baseboard and which can be adjusted so that the required sizes of disc may be cut. Your design should include some method of locating the blanks on the jig. (*AEB Craftwork – Design 'O'.*)

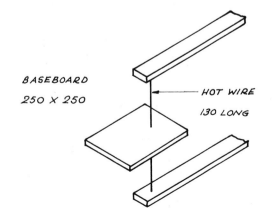

21 Design a piece of apparatus suitable for cutting the expanded polystyrene as in Question 20. This apparatus is to operate on a 6 volt supply and candidates' examination pieces will be treated at 6 volts but the power source which you use will not be required.

Note. During the period between the Design and Realisation of Design Examinations you are required to investigate:

(a) the melting point of polystyrene;

(b) the kind of wire which will produce the required heat for the purpose.

Attach to your Realisation of Design piece a record of your investigations. (*AEB Craftwork – Design 'O'.*)

22 The main entrance to a bungalow leads off a covered car port as shown.

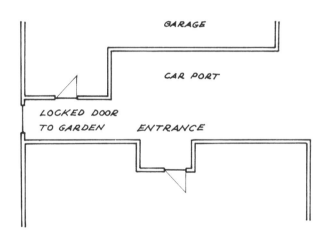

Design a device to give audible warning inside the bungalow of anyone entering the car port.

Make complete working drawings and a material list.

Include preliminary sketches leading up to your final design, together with any explanatory notes. Marks will be awarded for the development of ideas in sketch form.

Candidates are expected to show by their solutions to this problem an appreciation of materials, processes and construction. (*AEB Craftwork – Design 'O'.*)

23 Design a jig to be used for making skids for children's sledges.

Each skid is to be of 25 mm square section with curved fronts as shown and laminated from strips of ash.

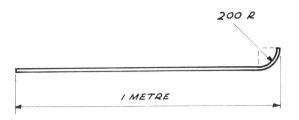

Indicate how your jig would be used. (*AEB Craftwork – Communication and Application.*)

24 The diagram below shows the dimensions of a battery-operated clock movement.

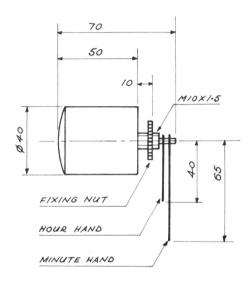

Design a suitable clock which incorporates this movement for use either:

    (a) on a wall of a lounge,

or **(b)** on the wall of a kitchen,

or **(c)** on a shelf in the lounge.

(*Note.* You must indicate on your drawing which of these three situations you have chosen.) (*AEB Craftwork – Design.*)

25 Design a complete lighting unit, suitable for use in ONE of the following situations:

(a) to be attached to a bedroom wall or bed-head in a suitable position for reading by someone in a single bed but such that it would not disturb someone else wishing to go to sleep in another bed in the same room;

131

(b) to be used on a table in a living room and capable of adjustment to give either background illumination to the whole room or, at other times, a more directional and concentrated illumination.

(*Note.* If exposed metal is used in your solution to the problem, you must indicate how you propose to 'earth' the metal for safety.) (*AEB Craftwork – Design.*)

26 Both outer doors of a house are to be fitted with bell pushes operating a single bell. Two indicators are to be installed to show which bell has been pressed. Each indicator should operate for a short time after the circuit has been broken.

Design a suitable circuit and indicator for this purpose to be operated by a $4\frac{1}{2}$-volt dry battery.

In the Realisation of Design Examination you will be required to make only one indicator. Commercially made components must *not* be used.

Your indicator must then be mounted on a board with one bell push, a $4\frac{1}{2}$-volt dry battery and a $4\frac{1}{2}$-volt bulb in place of the bell.

Only one circuit will be used but your indicator should operate when the bell push is pressed. (*AEB Craftwork – Design.*)

27 A balance is required to weigh small parcels *without* the use of a set of weights. It should measure in metric units in steps of 20 grams up to a maximum of 500 grams. Design a balance suitable for this purpose. (*AEB Craftwork – Design.*)

28 The diagram below shows in outline a glass-reinforced plastic body for a toy wheelbarrow.

(a) Show by means of sketches how you would make a suitable mould, to produce a number of

such bodies, and list the materials you would require.

(b) Draw a flow diagram to illustrate the processes and materials used in making one G.R.P. body from this mould. (*AEB Craftwork – Design.*)

29 A number of specimen plants grown in pots 250 mm diameter and 250 mm high are to be displayed, in succession, on a terrace outside a picture window.

Using materials of your own choice, design a container which will hide the pot and give greater stability.

Draw a flow diagram to illustrate how you proceed to make the container you have designed. (*AEB Craftwork – Design.*)

30 The diagram below shows a stator stamping for a small electric motor. When purchased these blanks do not have the 3 mm holes for clamping purposes.

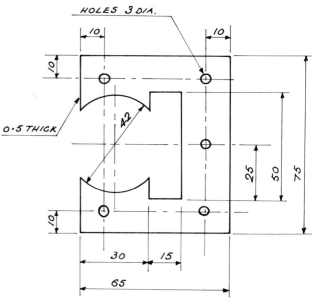

ALL DIMENSIONS IN MILLIMETRES

A large number of these have to be drilled in the positions shown in the drawing. It is suggested that these could be done in batches of 20.

Design a jig for this purpose.

Indicate the essential points in your design which will ensure accuracy in drilling and a uniform assembly of any batch of stampings. (*AEB Craftwork – Design.*)

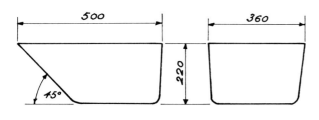

31 One sheet of 3 mm plywood measuring 2 metres by 1 metre is available.

Note. 3 mm plywood will give greatly increased strength when bent or laminated.

Design a toy suitable for a child of 6 years of age using the whole or part of this sheet.

Only a minimum use of other materials is permitted. (*AEB Craftwork – Design.*)

32 Your father wishes to build a combined garage and workshop attached to the house. In addition to the space occupied by the car a minimum of 9 square metres of workshop space is required.

The diagram right shows the plan of the part of the house adjacent to the boundary fence. The relevant doors and windows in that part of the house are shown.

The car is 3.5 metres long and 1.5 metres wide.

In addition to the car the following equipment is to be included:

one bench 1.5 metres long×0.75 metres deep;
one sink 1 metre long×0.5 metres deep;
one cupboard 1.5 metres long×0.5 metres deep;
one drill occupying a space of 0.5 metres×0.5 metres;
one lathe occupying a space of 1 metre×0.5 metres.

Draw a plan of the garage and workshop you would propose and show the positions of each of these items together with doors, windows, lights, power points and a place that must be provided for the dustbin so that it can be collected for emptying from the front of the house. (*AEB Craftwork – Design.*)

33 A simple light-operated mechanism is required which will ring a bell when a beam of light focused on it, is interrupted.

Sketch a suitable circuit and specify the components.

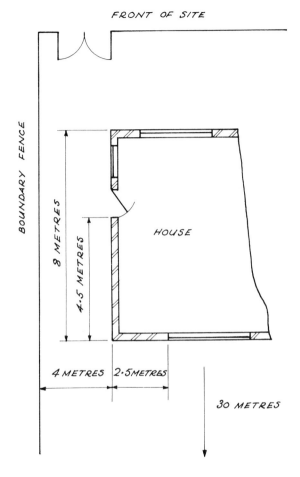

FRONT OF SITE

BOUNDARY FENCE

8 METRES

4.5 METRES

HOUSE

4 METRES 2.5METRES

30 METRES

Show, by means of sketches, how the apparatus should be set up to give warning that a customer (or unauthorized intruder) has entered a shop. (*AEB Craftwork – Design.*)

Appendix: Plastics in Schools Safety and Hazards*

This appendix contains advice on the handling of some of the solvents and other chemicals that are likely to be encountered during the study of plastics in schools. It also contains suggestions which it is hoped will assist in the safe and smooth operation of work areas in which plastics are used. It is not claimed to be a complete treatise on the handling of dangerous chemicals.

The use of plastics in various forms in schools, not only in the laboratories but also in workshops, art rooms and general classrooms, can introduce the possibility of new hazards arising in manipulation and storage. The range of plastics materials available for use need not be restricted unnecessarily if safe methods of handling are encouraged.

If in doubt about the hazards associated with any chemical, it is advisable to regard it as potentially hazardous and take precautions accordingly, for example work with it only in a fume cupboard away from sources of ignition and prevent it coming into contact with the skin. Useful advice on hazards is often provided by manufacturers, sometimes for instance on containers of chemicals, and it is advisable to note such advice carefully and to act in accordance with it.

RECOGNITION OF DANGEROUS LEVELS

The majority of the chemicals described in this appendix have recognizable odours and it is good practice to ensure that the concentration of any vapour is never allowed to build up to a level at which its odour is identifiable in the laboratory or craft room atmosphere.

Work with plastics, as with aerosols, paints and sprays, should not begin unless there is adequate ventilation, ideally in a fume cupboard. The ventilation should be sufficient to maintain a supply of fresh air in the work area at all times. It is suggested that six to eight air changes per hour is the minimum level at which to aim.

Because individuals differ in their susceptibility to some chemicals, serious attention should always be given to anyone who complains of headache or nausea. Such an individual should be encouraged and helped to seek fresh air. Individuals susceptible to dermatitis should be identified by the teacher and advised privately to wear protective gloves and barrier creams.

Personal protection

It is necessary to stress the value of protective clothing in the laboratory or craft room. The importance of wearing safety spectacles or face visors in any situation where chemicals may be splashed into the eyes, or where there is a risk of dust or small particles entering the eyes, must be emphasized. In situations involving the handling of hot materials the wearing of suitable, dry, protective gloves is essential. Open cuts and abrasions should be suitably protected before work is started.

Fire and explosion hazards

a All plastics materials should be stored under cold, dry conditions, preferably in an outside, brick-built, store-room. The storage of large quantities of catalysts, resins and cleaning fluids will increase the fire risk and can also be wasteful because of the limited shelf life of some materials. A three-month stock may, in general, be considered to be sufficient, economic and safe.

Organic peroxides (catalysts) should be stored in vented containers in cool, separate, preferably metal, cabinets, away from flammable materials. Catalysts and accelerators should never be mixed directly together because this may produce a violent and possibly explosive reaction. The use of pre-accelerated polyester resins removes this particular risk.

---

* This appendix is reproduced by courtesy of The Plastics and Rubber Institute from their booklet *Plastics in Schools – Safety and Hazards.*

b The disposal of large quantities of waste plastics materials is, of course, a matter for the specialist. In schools only small quantities are likely to be involved. However, waste material should not be left in the work area but should be removed and disposed of as soon as possible.

Small quantities of liquid waste (usually solvents) can be disposed of by allowing such waste to evaporate in the open air from a metal tray in an area free from fire risk and not accessible to children.

Solid material (e.g. surplus cured resin, machine swarf) may be placed in a labelled polyethylene sack and stored outside for collection by the local authority. Surplus catalysed resin should be spread thinly on a metal tray and allowed to harden before disposal.

Rags which have been used to mop up spillage should be placed in a metal bin outside the building and burnt as soon as possible. Rags soaked in peroxide solutions must *not* be mixed with other waste material.

c Foamed plastics materials, because of their high surface area, are highly flammable and must be stored with care away from heat sources, open flames and other sources of ignition. Some grades of foamed materials are compounded to provide a measure of fire retardance. These, however, will still burn vigorously once they have been ignited, but they are more difficult to ignite than non-flame-retardant grades.

HEALTH HAZARDS AND FIRST AID

Prevention of accidents is always preferable to putting things right after an accident has occurred. 'Accident' is perhaps the wrong word to use here as it suggests that something has happened by chance. Dangerous occurrences in laboratories do not happen by chance, they happen because someone has not been properly instructed or has not followed his instructions correctly. However, a First Aid box equipped to deal with burns and cuts and for eye irrigation, should always be provided in any area where plastics materials are used.

The First Aid actions to be taken after an incident will depend on the type of contact the patient has had with a chemical or the type of injury he has suffered. If qualified medical assistance can be obtained quickly enough it is probably best merely to remove the patient from the source of danger

and make him comfortable, leaving the treatment to a qualified person. However, such a course of action is not always possible and therefore, after discussion of the relevant problems, treatments are described below which will provide some measure of First Aid relief to the patient. Qualified medical advice, however, should always be obtained as soon as possible after any incident.

Inhalation

Harmful gases or vapours may be produced through the evaporation of solvents and cements and by the breakdown of materials through the application of heat, and it is necessary that their concentrations are kept as low as possible. Inhalation of toxic vapours may have delayed reactions so that danger is not immediately apparent.

a The production of articles in glass-reinforced polyester (GRP) results in the liberation of phenylethene (styrene) fumes into the air. Provided that only small quantities are being used in any one area, and that the area is well ventilated, the phenylethene fumes are unlikely to reach a harmful concentration (TLV 100 parts per million). The odour of phenylethene is such that it is normally obvious well below this level. If larger mouldings are being produced then some form of forced ventilation is needed, and when it is necessary to work inside such a moulding, e.g. in the moulding of boathulls, a supply of fresh air to the inside is required as well as the extraction of fumes. It should be remembered that in hot weather the concentration of phenylethene and solvent vapour in the atmosphere is likely to be higher due to the increased evaporation and under these conditions adequate ventilation of the work area is essential.

b Cutting expanded poly (phenylethene) (polystyrene) by means of a hot wire generates phenylethene fumes. This should be carried out only in well ventilated conditions so that only a low concentration of vapour is obtained. Phenylethene fumes may irritate the eyes and cause dizziness if concentrated, although this depends to some extent on personal susceptibility. Children appear to be more prone to this hazard than do adults and it is therefore recommended that ventilation should be generous in every case. The cutter should be so constructed as to operate at an even temperature below red heat. If smoke is given off then the cutting wire is too hot. An even wire temperature cannot easily be obtained without the use of controlled electrical heating (see also *General safety notes*).

c When casting metals using expanded poly (phenylethene) patterns, it is essential that good venting of the mould be provided in order to minimize the risk of an accumulation of vapour, and that fumes be exhausted directly to the outside air.

d The production of polyurethane foam from liquid materials produces toxic gases and it is advised that this procedure should be carried on in the open and under the *direct* control of an *experienced* adult. *Compositions based on toluene diisocyanate (TDI) should never be used in schools or colleges.* In general, the materials for manufacturing rigid polyurethane foams present less of a hazard than do those used for flexible foams.

e Degreasing should not be carried out with solvents such as petrol, ethanol, ethers or ketones. Not only are these flammable; some are also highly toxic. Tetrachloromethane (carbon tetrachloride), whilst it is non-flammable, is highly toxic and should not be used. Appropriate detergents may be used, providing suitable protective gloves are worn.

f Precautions should be taken when machining or abrading all materials to prevent excessive inhalation of dust and small particles. Like some timbers and ceramics, many plastics are inherently dusty. Adequate ventilation is essential and inexpensive disposable masks should be provided. Hand rather than machine methods of abrading are advised, especially when water can be used as a lubricant, because in this way dust production can be kept to a minimum. Dust produced on work with GRP can be reduced if the material is trimmed whilst it is in the 'green' stage.

First Aid

Remove the patient from the source of vapour to the open air.
Make him rest and keep him warm.
Make him lie down with his feet slightly raised, with his clothing loosened but with a blanket placed lightly over him if necessary.
If breathing has stopped apply mouth-to-mouth resuscitation.
Mild shock can be relieved with coffee or tea but do *not* offer alcohol except under medical advice, which should be obtained as soon as possible.

Eye contact

Apart from dust the eyes are at risk from waste particles of plastics and from solvent and cement splashes.
Organic peroxides which are used as catalysts for curing GRP will cause severe damage if in contact with the eye. Measuring or mixing of the catalyst should be done only by the teacher or under his direct supervision. A standard dispenser which is constructed to prevent squirting of the liquid and which is calibrated to allow accurate measurement should be used.

First Aid

Irrigate the eye immediately and copiously.
Obtain medical advice as soon as possible.

Skin contact

a Dermatitis risk is present with the use of resins and adhesives, especially with epoxy and polyurethane materials, which should only be used under strict control. Good washing facilities should be available and any resin contamination should be washed off at once when using uncured epoxy and polyurethane resins. Exposed areas of the skin should be protected with a barrier cream before starting work and suitable protective gloves should be worn. The use of a proprietary cream to cleanse the skin is prudent, but the use of solvents, such as propanone (acetone) for this purpose should be avoided because frequent degreasing can result in harmful effects to the body.

First Aid

Drench the skin with water and remove contaminated clothing.
Soap detergent will facilitate the removal of substances insoluble in water.

b The temperatures involved in plastics moulding and shaping are lower than those for comparable metal processes. Nevertheless, attention must be drawn to the dangers of allowing molten plastics to come into contact with the skin. These dangers arise from the high heat capacity of molten plastics and from the fact that they stick to the skin and are therefore difficult to remove. Protection can usually be afforded by the use of thick, dry, industrial gloves.
Injection moulding machines do present the hazard of accidental ejection of molten plastics and adequate guarding of such machines is essential.

First Aid

Dry dressings are recommended for heat and chemical burns; they can be held in place by cotton wool pads secured with a bandage or adhesive tape. Such dressings should only be regarded as offering protection until medical advice is obtained.

These treatments are First Aid actions to prevent further injury to the patient. Where it is necessary to carry out further treatment as the result of contact with particular chemicals, this is indicated in the table overleaf and it is advisable to consult this table as soon as First Aid has been initiated. If in doubt, call in qualified medical assistance as soon as possible and *always* consult a doctor if chemicals have been swallowed or have been splashed in the eyes.

A list of the chemicals which may be needed in a school in which work involving polymers and plastics is done is provided in the table together with data on their hazards and ways of overcoming these.

GENERAL SAFETY NOTES

a All electrical connections should conform to the standards laid down in the latest edition of the Institution of Electrical Engineers Wiring regulations, and should be made only by a competent electrician. School-made equipment containing electrical components should always be tested by a recognized competent person before it is brought into use. Periodic testing (e.g. for earth continuity) should also be carried out.

b Electrical power for hot wire cutters should be supplied preferably by batteries rather than from the mains supply unless an approved isolating low voltage transformer is available.

c A chemical trap for the sink is advised as the sink is likely to be used for washing-off resins and other chemicals. Accumulation of materials does not then occur in the drains. Solvents and other waste materials that are immiscible with water should *not* be put down the sink.

Whilst the information in this appendix is given in good faith the Plastics and Rubber Institute, the author and publishers can accept no liability arising out of its use.

The booklet from which this appendix is taken has been designed so that, by utilizing two copies, it may be used as a wall-chart.

Copies of the booklet, 'Plastics in Schools – Safety and Hazards', may be obtained, price 25p each (cash with order please), from the Plastics and Rubber Institute, 11 Hobart Place, London SW1W 0HL.

Education authorities or other organizations interested in bulk supplies should contact the Education Officer at the above address (tel. 01-245 9555).

Some chemicals likely to be encountered, their hazards and the precautions for safe working

Substance		TLV (in ppm)	Hazards	Precautions and further actions
IUPAC Systematic name	Common name			
Ethanoic acid	**Acetic acid**	10	Irritant vapour, corrosive, causes burns. Flammable.	Avoid breathing vapour, avoid contact with eyes and skin. If swallowed wash mouth thoroughly with water and give water to drink, followed by Milk of Magnesia as an emetic.
Ethanoic anhydride	**Acetic anhydride**	5	Irritant vapour, corrosive, causes burns. Flammable.	As for acetic acid.
Propanone	**Acetone**	1000	Highly flammable, vapour/air mixture explosive. Harmful vapour.	Avoid breathing vapour; avoid contact with eyes. Ventilate clothing onto which propanone has been spilt.
Aluminium chloride	**Aluminium chloride**		Harmful dust, corrosive, causes burns.	Avoid breathing dust, prevent contact with skin or eyes. If swallowed give plenty of water to drink, followed by Milk of Magnesia.
Ammonia	**Ammonia**	100	Irritant vapour, corrosive, causes burns. Harmful if taken by mouth.	Avoid breathing vapour, prevent contact with eyes or skin. If swallowed give plenty of water followed by vinegar (or 1 per cent acetic acid) to drink.
Benzene	**Benzene**	25	Harmful vapour, harmful by skin absorption. Highly flammable.	Avoid inhalation of vapour, prevent contact with skin and eyes. If swallowed give an emetic.
Dibenzoyl peroxide	**Benzoyl peroxide**		Explosive when dry, spontaneously explosive when mixed with accelerator without resin.	Avoid contact with skin and eyes. Dibenzoyl peroxide can usually be replaced by di (dodecanoyl) peroxide (lauroyl peroxide) which is less sensitive to detonation.
Butyl 2-methyl-propenoate	**Butyl methacrylate**		Harmful vapour, harmful by skin absorption. Flammable.	Avoid breathing vapour, avoid contact with skin and eyes.
Tetrachloro-methane	**Carbon tetrachloride**	10	Harmful vapour.	Avoid breathing vapour, avoid contact with skin and eyes.
1-Chloro-2, 3-epoxy-propane	**Epichlorhydrin**	5	Harmful, irritant vapour, harmful by skin absorption. Flammable.	Avoid breathing vapour, avoid contact with skin and eyes. If swallowed give an emetic.
1,6-diamino-hexane	**Hexamethylene diamine**		Irritant vapour, harmful by skin absorption. Flammable.	Avoid breathing vapour, avoid contact with skin and eyes. If swallowed wash out mouth thoroughly with water and give water to drink.
1,2-dichloro-ethane	**Ethylene dichloride**	50	Harmful vapour. Highly flammable.	Avoid breathing vapour, avoid contact with skin and eyes. If swallowed give an emetic.
Methanal solution	**Formaldehyde solution (formalin)**	5	Irritant vapour.	Avoid breathing vapour, avoid contact with skin and eyes. If swallowed give milk freely to drink.
Furanol	**Furfuryl alcohol**		Harmful vapour. Flammable.	Avoid breathing vapour.
Hydrogen chloride solution	**Hydrochloric acid**	5	Irritant vapour. Corrosive, causes burns.	Avoid breathing vapour, avoid contact with skin and eyes. If swallowed give water followed by Milk of Magnesia.

Substance		TLV (in ppm)	Hazards	Precautions and further actions
IUPAC Systematic name	Common name			
Di(dodeca-noyl) peroxide	Lauroyl peroxide		Flammable. Oxidizing agent, assists fire.	Avoid contact with skin, eyes and clothing.
Methyl 2-methyl-propenoate	Methyl methacrylate	100	Harmful vapour. Highly flammable.	Avoid breathing vapour, avoid contact with skin and eyes.
Phenol	Phenol	5	Harmful vapour, harmful by skin absorption. Corrosive, causes burns.	Avoid breathing vapour, avoid contact with skin and eyes. If swallowed induce vomiting by placing a finger far back into throat; summon medical attention immediately and after vomiting has ceased wash mouth out thoroughly with water.
	Polyester resin			Treat as for phenylethene (styrene).
Benzene-1,3-diol	Resorcinol		Harmful by skin absorption.	Avoid contact with skin and eyes. If swallowed treat as for phenol.
Decane dioyl chloride	Sebacoyl chloride		Irritant vapour. Corrosive, causes burns. Flammable.	Avoid breathing vapour, avoid contact with skin and eyes. If swallowed give plenty of water to drink followed by Milk of Magnesia.
Phenylethene	Styrene	100	Irritant vapour. Flammable.	Avoid breathing vapour, avoid contact with skin and eyes. If swallowed give an emetic.
Methyl-benzene	Toluene	200	Harmful vapour, harmful by skin absorption. Highly flammable.	Avoid breathing vapour, avoid contact with skin and eyes.
1,1,2-tri-chloroethene	Trichloro-ethylene	100	Harmful vapour.	Avoid breathing vapour, avoid contact with skin and eyes.
1,1,1-tri-chloroethane	Trichloro-ethane	350	Harmful vapour.	Avoid breathing vapour, avoid contact with skin, eyes and clothing.
Ethenyl ethanoate	Vinyl acetate		Highly flammable.	Avoid breathing vapour, avoid contact with skin and eyes.

The information in the above table is given as a guide to help in safe working and to assist if an accident does occur. It must be stressed that qualified medical advice should be obtained as soon as possible after an accident, particularly if it is suspected that dust or chemicals have been swallowed or have entered the eyes. It is better to be over cautious with unfamiliar materials than to take unnecessary risks.

Index